VOLUME SIXTY FOUR

ADVANCES IN MICROBIAL PHYSIOLOGY

Advances in Microbial Systems Biology

VOLUME SIXTY FOUR

ADVANCES IN MICROBIAL PHYSIOLOGY

Advances in Microbial Systems Biology

Edited by

ROBERT K. POOLE
West Riding Professor of Microbiology
Department of Molecular Biology and Biotechnology
The University of Sheffield
Firth Court, Western Bank
Sheffield, UK

ELSEVIER

AMSTERDAM • BOSTON • HEIDELBERG • LONDON
NEW YORK • OXFORD • PARIS • SAN DIEGO
SAN FRANCISCO • SINGAPORE • SYDNEY • TOKYO
Academic Press is an imprint of Elsevier

Academic Press is an imprint of Elsevier
32 Jamestown Road, London NW1 7BY, UK
The Boulevard, Langford Lane, Kidlington, Oxford, OX5 1GB, UK
Radarweg 29, PO Box 211, 1000 AE Amsterdam, The Netherlands
225 Wyman Street, Waltham, MA 02451, USA
525 B Street, Suite 1800, San Diego, CA 92101-4495, USA

First edition 2014

ISBN: 978-0-12-800143-1
ISSN: 0065-2911

CONTENTS

CONTRIBUTORS

Fiona Achcar
Institute of Molecular, Cell and Systems Biology, College of Medical, Veterinary and Life Sciences, University of Glasgow, Glasgow, United Kingdom

Joaquín Ariño
Institut de Biotecnologia i Biomedicina & Dept. Bioquímica i Biologia Molecular, Universitat Autònoma de Barcelona, Cerdanyola del Vallès, Barcelona, Spain

Ebru Aydar
Faculty of Life Sciences and Computing, London Metropolitan University, London, United Kingdom

Hao Bai
Department of Computer Science, The University of Sheffield, Sheffield, United Kingdom

Barbara M. Bakker
Department of Pediatrics, Centre for Liver Digestive and Metabolic Diseases, and Systems Biology Centre for Energy Metabolism and Ageing, University Medical Center Groningen, University of Groningen, Groningen, The Netherlands

Michael P. Barrett
Wellcome Trust Centre for Molecular Parasitology, Institute of Infection, Immunity and Inflammation, and Glasgow Polyomics, College of Medical, Veterinary and Life Sciences, University of Glasgow, Glasgow, United Kingdom

Katja Bettenbrock
Max Planck Institute for Dynamics of Complex Technical Systems, Magdeburg, Germany

Rainer Breitling
Manchester Institute of Biotechnology, Faculty of Life Sciences, University of Manchester, Manchester, United Kingdom

Christine Clayton
Zentrum für Molekulare Biologie der Universität Heidelberg, DKFZ-ZMBH Alliance, Heidelberg, Germany

Samuel Drulhe
Humboldt University, Berlin, Germany

Michael Ederer
Institute for System Dynamics, University of Stuttgart, Stuttgart, Germany

Abeer Fadda
Zentrum für Molekulare Biologie der Universität Heidelberg, DKFZ-ZMBH Alliance, Heidelberg, Germany

Daniel Ganser
University Bonn, Bonn, Germany

Mark Girolami
The Department of Statistical Science and The Centre for Computational Statistics and Machine Learning University College London, London, United Kingdom

Jeffrey Green
Department of Molecular Biology and Biotechnology, The University of Sheffield, Sheffield, United Kingdom

Jurgen R. Haanstra
Department of Pediatrics, Centre for Liver Digestive and Metabolic Diseases, and Systems Biology Centre for Energy Metabolism and Ageing, University Medical Center Groningen, University of Groningen, Groningen, and Department of Molecular Cell Physiology, Faculty of Earth and Life Sciences, VU University Amsterdam, Amsterdam, The Netherlands

Klaas J. Hellingwerf
Swammerdam Institute for Life Sciences, University of Amsterdam, Amsterdam, The Netherlands

Michael Holcombe
Department of Computer Science, The University of Sheffield, Sheffield, United Kingdom

Jesús Jorrín
Department of Biochemistry and Molecular Biology, University of Córdoba, Córdoba, Spain

Matthias Kahm
RheinAhrCampus, University of Applied Sciences, Koblenz, Remagen, Germany

Eduard J. Kerkhoven
Wellcome Trust Centre for Molecular Parasitology, Institute of Infection, Immunity and Inflammation, and Glasgow Polyomics, College of Medical, Veterinary and Life Sciences, University of Glasgow, Glasgow, United Kingdom, and Systems and Synthetic Biology Group, Department of Chemical and Biological Engineering, Chalmers University of Technology, Gothenburg, Sweden

Dong-Hyun Kim
Wellcome Trust Centre for Molecular Parasitology, Institute of Infection, Immunity and Inflammation, and Glasgow Polyomics, College of Medical, Veterinary and Life Sciences, University of Glasgow, Glasgow, United Kingdom

Falko Krause
Humboldt University, Berlin, Germany

R. Luise Krauth-Siegel
Biochemie-Zentrum der Universität Heidelberg, Heidelberg, Germany

Maik Kschischo
RheinAhrCampus, University of Applied Sciences Koblenz, Remagen, Germany

Samantha Kunz
Institute for System Dynamics, University of Stuttgart, Stuttgart, Germany

Alejandro E. Leroux
Biochemie-Zentrum der Universität Heidelberg, Heidelberg, Germany

Jost Ludwig
University Bonn, Bonn, Germany

Keith R. Matthews
Centre for Immunity, Infection and Evolution, Institute for Immunology and Infection Research, School of Biological Sciences, Ashworth Laboratories, University of Edinburgh, Edinburgh, United Kingdom

Chris Palmer
Faculty of Life Sciences and Computing, London Metropolitan University, London, United Kingdom

Theodore Papamarkou
The Department of Statistical Science and The Centre for Computational Statistics and Machine Learning University College London, London, United Kingdom

Silvia Petrezsélyová
Institut de Biotecnologia i Biomedicina & Dept. Bioquímica i Biologia Molecular, Universitat Autònoma de Barcelona, Cerdanyola del Vallès, Barcelona, Spain

Robert K. Poole
Department of Molecular Biology and Biotechnology, The University of Sheffield, Sheffield, United Kingdom

José Ramos
Department of Microbiology, University of Córdoba, Córdoba, Spain

Federico Rojas
Centre for Immunity, Infection and Evolution, Institute for Immunology and Infection Research, School of Biological Sciences, Ashworth Laboratories, University of Edinburgh, Edinburgh, United Kingdom

Matthew D. Rolfe
Department of Molecular Biology and Biotechnology, The University of Sheffield, Sheffield, United Kingdom

Guido Sanguinetti
School of Informatics, University of Edinburgh, Edinburgh, United Kingdom

Oliver Sawodny
Institute for System Dynamics, University of Stuttgart, Stuttgart, Germany

Poonam Sharma
Department of Molecular Biology and Biotechnology, The University of Sheffield, Sheffield, United Kingdom

Sonja Steinsiek
Max Planck Institute for Dynamics of Complex Technical Systems, Magdeburg, Germany

Hana Sychrová
Department of Membrane Transport, Institute of Physiology Academy of Sciences CR, Prague, Czech Republic

G. Paul H. van Heusden
Institute of Biology, Leiden University, Leiden, The Netherlands

Lynne Yenush
Instituto de Biología Molecular y Celular de Plantas, Universidad Politécnica de Valencia-Consejo Superior de Investigaciones Científicas, Valencia, Spain

Olga Zimmermannová
Department of Membrane Transport, Institute of Physiology Academy of Sciences CR, Prague, Czech Republic

PREFACE

Systems biology is an approach to the study of biology that is clearly distinguishable from the 'gene's eye' view and considers the cell or organism as a complex system of many interacting strata. Although a biology-based field of study, its focus on complex interactions requires a holistic perspective instead of the more traditional reductionist approach to biological and biomedical research. It is as Denis Noble has written 'Biology beyond the genome' (Noble, 2006). One of the distinctive aims of systems biology is to model and discover emergent properties, that is, features of the system whose theoretical description involves metabolic or signalling networks. Consequently, systems biology makes heavy use of mathematical, statistical and computational models and may inspire new mathematical theories or generate new generally applicable models.

This volume brings together three programmes of research that contributed to the SysMO Consortium. SysMO (Systems Biology of Microorganisms) was a European transnational funding initiative that aimed to record and describe the dynamic molecular processes occurring in microorganisms and to present these processes in the form of mathematical models. Each of the individual projects in SysMO worked towards different research outcomes and represented a cross-section of microorganisms, including bacteria, archaea and yeast. The environmental conditions for each organism also varied widely with organisms growing in laboratory culture, soil, water and animal hosts. In this volume are collected three accounts of SysMO-funded research on yeast (the Translucent and Translucent-2 projects on gene interaction networks and models of cation homeostasis in *Saccharomyces cerevisiae*), SUMO and SUMO-2 (Systems Understanding of Microbial Oxygen Responses, as applied to *Escherichia coli*) and SilicoTryp (the creation of a 'Silicon Trypanosome', a comprehensive, experiment-based, multi-scale mathematical model of trypanosome physiology).

It should be clear from this small selection of projects that, intrinsic to systems biology, is its interdisciplinary nature and the common aim of achieving the quantitative understanding of dynamic biological processes through the use of mathematical and statistical analyses. As a consequence of this diversity, there is no one model for experimentation or for the types of data collected or the types of models produced. In order to pool the research outcomes for SysMO, the SysMO-DB programme (http://www.sysmo-db.org)

was established to support and manage this diversity and promote a shared understanding across the community by using the same technologies. SysMO-DB has created a Web-based platform and tools for finding, sharing and exchanging data, models and processes in SysMO, but the principles and methods employed are equally applicable to other multi-site Systems Biology projects. Further details on these three projects (the scientific partners and their funding bodies in the participating countries) as well as the several other projects that could not be presented can here be found at that address.

REFERENCE

Noble, D. (2006). *The music of life.* Oxford: Oxford University Press.

CHAPTER ONE

Systems Biology of Monovalent Cation Homeostasis in Yeast: The Translucent Contribution

Joaquín Ariño[*,1], Ebru Aydar[†], Samuel Drulhe[‡], Daniel Ganser[§], Jesús Jorrín[¶], Matthias Kahm[#], Falko Krause[‡], Silvia Petrezsélyová[*], Lynne Yenush[**], Olga Zimmermannová[††], G. Paul H. van Heusden[‡‡], Maik Kschischo[#], Jost Ludwig[§], Chris Palmer[†], José Ramos[§§], Hana Sychrová[††]

[*]Institut de Biotecnologia i Biomedicina & Dept. Bioquímica i Biologia Molecular, Universitat Autònoma de Barcelona, Cerdanyola del Vallès, Barcelona, Spain
[†]Faculty of Life Sciences and Computing, London Metropolitan University, London, United Kingdom
[‡]Humboldt University, Berlin, Germany
[§]University Bonn, Bonn, Germany
[¶]Department of Biochemistry and Molecular Biology, University of Córdoba, Córdoba, Spain
[#]RheinAhrCampus, University of Applied Sciences Koblenz, Remagen, Germany
[**]Instituto de Biología Molecular y Celular de Plantas, Universidad Politécnica de Valencia-Consejo Superior de Investigaciones Científicas, Valencia, Spain
[††]Department of Membrane Transport, Institute of Physiology Academy of Sciences CR, Prague, Czech Republic
[‡‡]Institute of Biology, Leiden University, Leiden, The Netherlands
[§§]Department of Microbiology, University of Córdoba, Córdoba, Spain
[1]Corresponding author: e-mail address: Joaquin.Arino@uab.es

Contents

Advances in Microbial Physiology, Volume 64
ISSN 0065-2911
http://dx.doi.org/10.1016/B978-0-12-800143-1.00001-4

Abstract

Maintenance of monovalent cation homeostasis (mainly K^+ and Na^+) is vital for cell survival, and cation toxicity is at the basis of a myriad of relevant phenomena, such as salt stress in crops and diverse human diseases. Full understanding of the importance of monovalent cations in the biology of the cell can only be achieved from a systemic perspective. Translucent is a multinational project developed within the context of the SysMO (System Biology of Microorganisms) initiative and focussed in the study of cation homeostasis using the well-known yeast *Saccharomyces cerevisiae* as a model. The present review summarize how the combination of biochemical, genetic, genomic and computational approaches has boosted our knowledge in this field, providing the basis for a more comprehensive and coherent vision of the role of monovalent cations in the biology of the cell.

ABBREVIATIONS

2-DE two-dimensional IEF-SDS-PAGE electrophoresis
AAS atomic absorption spectrophotometry
BiFC bimolecular fluorescence complementation
CDRE calcineurin-dependent response element
ChiP chromatin immunoprecipitation
diS-C_3(3) 3,3′-dipropylthiadicarbocyanine
GFP green fluorescent protein
HA hemagglutinin
IEF isoelectrofocusing
MS mass spectrometry
PAGE polyacrylamide gel electrophoresis
PKA protein kinase A
ROS reactive oxygen species

1. TRANSLUCENT 1 AND 2, A STORY OF COLLABORATIVE LEARNING

The Translucent Project (Translucent 1) was initiated in the context of the SysMO initiative, a European transnational funding and research effort on "Systems Biology of Microorganisms" (http://www.sysmo.eu/), as a result of a funding call issued in 2005. Subsequently, a second call appeared in 2009. The Translucent proposal (Translucent 2) was selected again for funding in a tight competition with many other proposals (only 7 research projects from out of a total of 45 proposals were funded). This allowed us to capitalize on all the tools and experience gained in the Translucent 1 project and to further expand modelling of cation homeostasis in yeast.

Understanding of the mechanisms of monovalent cation homeostasis (mainly K^+ and Na^+) is crucial for the interpretation of many other biological events, as these cations define diverse cellular parameters, such as membrane potential, cellular turgor, intracellular pH, etc. In addition, an imbalance of cellular K^+ and Na^+ ions might be lethal for microorganisms that cannot cope with sudden environmental changes; it has severe consequences for human health, as well as a detrimental effect on many agricultural crop species. At the initiation of Translucent, it was evident that the yeast *Saccharomyces cerevisiae* was an excellent model to understand the basis for such toxicity at a systemic level. Around 2005, most of the cation transporter/channel systems for sodium, potassium, calcium and diverse heavy metals were already identified in yeast. Similarly, the regulatory elements and control networks were relatively well defined. However, all this information provided only a static, often partial and, certainly, non-integrative view. Therefore, it seemed the right time for a comprehensive approach that could lead to quantitative understanding of this process in the form of mathematical models.

Most of the groups constituting Translucent 1 were also partners in Translucent 2 (see Table 1.1), with the only exception of University of Vienna, which could not participate in Translucent 2 because Austria was not a partner in the second SysMO call. The size of the consortium remained the same by including expert electrophysiologists from the London

Table 1.1 Composition of Translucent 1 (T1) and Translucent 2 (T2) consortia

Country	Affiliations	Group leader	T1	T2
Spain	Universitat Autònoma de Barcelona, Dept. of Biochemistry & Mol. Biol., – Institut de Biotecnologia i Biomedicina	Joaquín Ariño	✓	✓
Germany	Humboldt-Universität zu Berlin, Institut für Biologie, Theoretische Biophysik	Edda Klipp	✓	✓
Germany	Universität Bonn, Institut für Zelluläre und Molekulare Botanik – FH Koblenz, Rheinahrcampus Remagen	Jost Ludwig – Maik Kschischo	✓	✓
Spain	University of Córdoba, Department of Microbiology	José Ramos	✓	✓
The Netherlands	Institute of Biology, Leiden University, Leiden	Paul van Heusden	✓	✓
Austria	University of Vienna, Austria	Rudolf J. Schweyen	✓	
United Kingdom	London Metropolitan Univ., Institute for Health Research and Policy	Christopher P. Palmer		✓
Czech Republic (associated partner)	Department of Membrane Transport, Institute of Physiology AS CR	Hana Sychrová	✓	✓

Metropolitan University. The experience accumulated during this collaborative work involving diverse experimentalists and computational groups and lasting for more than 6 years has been enormous. This has been based on a very tight interaction, including general scientific meetings every 6 months, continuous exchange of scientists among groups, etc., and is reflected in the publication of numerous joint papers. During this period, we have experienced moments of great excitement, for instance when some hypothesis raised from the modelling could be confirmed or refined, or an important joint paper was accepted. We also had days of sadness, such as when in February 2009, we were told that our colleague and friend Rudolf J. Schweyen, group leader at the University of Vienna partner, had unexpectedly passed away.

In the following paragraphs, we describe the main advances generated by the Translucent 1 and 2 projects and how these findings provide a more comprehensive understanding of monovalent cation homeostasis and its interaction with the biology of the yeast cell.

2. OVERVIEW OF ALKALI-METAL-CATION TRANSPORTERS CONTRIBUTING TO POTASSIUM HOMEOSTASIS IN *S. CEREVISIAE*

The ability to maintain cation and pH homeostasis in changing environmental conditions is among the crucial requirements for all microorganisms. A high intracellular concentration of potassium is required for many physiological functions, such as regulation of cell volume, maintenance of stable potential across the plasma membrane, compensation of negative charges in many macromolecules, protein synthesis and enzyme activation (Arino, Ramos, & Sychrova, 2010; Cyert & Philpott, 2013). On the other hand, high intracellular concentrations of other alkali-metal-cations are toxic. Nevertheless, *S. cerevisiae* cells are able to grow in the presence of a broad range of external concentrations of K^+ (10 μM to 2.5 M) and Na^+ (<1.8 M) due to the presence of a variety of plasma membrane and organellar transporters that mediate the fluxes of cations with differing mechanisms and affinities (Arino et al., 2010). The plasma membrane of *S. cerevisiae* cells harbours two efficient potassium uptake uniporters, two types of alkali-metal-cation efflux systems (antiporter and ATPase), and at least one cation channel. In the membranes of organelles, five systems for transport of alkali-metal-cations have been identified so far. The activity of most potassium transporters depends on the gradient of protons and/or

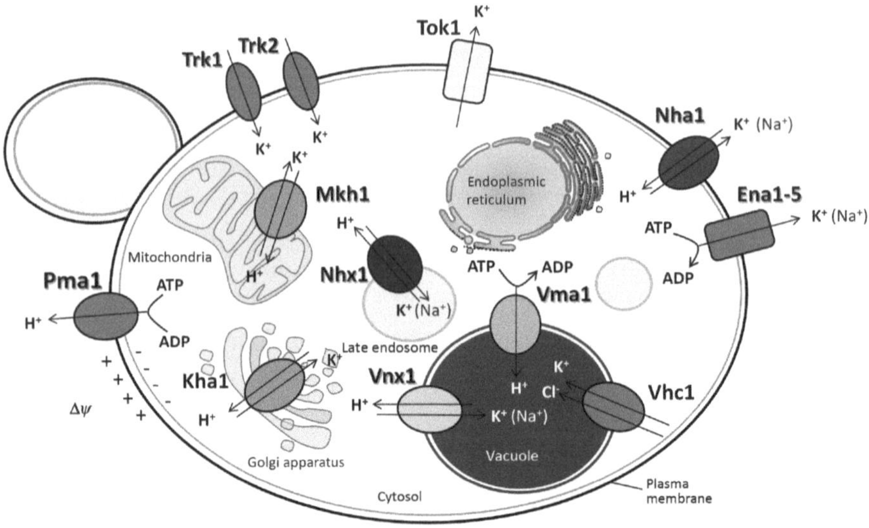

Figure 1.1 Plasma membrane and organellar *S. cerevisiae* transporters contributing to the potassium and pH homeostases differ in the mechanism of transport. Pma1 and Vma1, proton ATPases; Trk1 and Trk2, potassium uniporters; Nha1, Nhx1, Vnx1 and Kha1, cation/proton antiporters; Vhc1, cation-chloride cotransporter; Ena1-5, cation efflux ATPases; Tok1, potassium channel; and Mkh1, mitochondrial system exchanging potassium for protons.

potential across cell membranes, and thus their activity depends on the activity of the plasma membrane (Pma1) and vacuolar (Vma1) H^+ ATPases (Fig. 1.1).

2.1. Plasma membrane potassium transporters

TRK1 and *TRK2* genes encode partially redundant K^+ uptake systems (uniporters, Fig. 1.1) that are able to efficiently accumulate necessary potassium cations (Rodriguez-Navarro, 2000). Trk1 is the prominent K^+ uptake transporter and though Trk2 is much less effective, our studies showed that its activity contributes significantly to K^+ supply and the maintenance of plasma membrane potential (Petrezselyova, Ramos, & Sychrova, 2011). Recently, a new pair of putative low-affinity K^+ uptake systems, Kch1 and Kch2, that play a role in calcium signalling, has been partly characterized (Stefan et al., 2013).

Three systems mediate cation export. Tok1 is a potassium-specific outward rectifying channel (Bertl, Bihler, Reid, Kettner, & Slayman, 1998; Ketchum, Joiner, Sellers, Kaczmarek, & Goldstein, 1995) whose activity contributes significantly to the maintenance of the plasma membrane potential (Maresova, Urbankova, Gaskova, & Sychrova, 2006). The other two systems export both toxic cations (Na^+, Li^+, Rb^+) and surplus K^+ from cells; thus; they have a dual function—detoxification and maintenance of cation and pH homeostasis. Nha1 is an antiporter that uses the inward gradient of protons created by the Pma1 ATPase to actively pump out cations (Kinclova, Ramos, Potier, & Sychrova, 2001) and *ENA* genes (usually, in several tandem copies in the *S. cerevisiae* genome; Garciadeblas et al., 1993; Wieland, Nitsche, Strayle, Steiner, & Rudolph, 1995) encode ATPases pumping cations out of cells mainly at high external pH (Ruiz & Arino, 2007). The plasma membrane transporters tightly collaborate in a coordinated manner and our studies showed that the efficient and continuous efflux and influx of potassium cations is necessary for yeast cell fitness (cf. below; Navarrete et al., 2010; Zahradka & Sychrova, 2012).

2.2. Potassium transporters in organelles

Most of the intracellular organelles use systems with cation/proton antiport mechanisms to regulate the intraorganellar cation concentration and luminal pH, and to contribute to various other functions, for example, intracellular trafficking of vesicles (Brett, Tukaye, Mukherjee, & Rao, 2005). These antiporters function in the membranes of the Golgi apparatus (GA) (Kha1; Maresova & Sychrova, 2005), endosomes (Nhx1; Nass & Rao, 1998) and in vacuoles (Vnx1; Cagnac, Leterrier, Yeager, & Blumwald, 2007). The sequestration of surplus cations into vacuoles is very important for yeast physiology, and in addition to the two antiporters mediating the accumulation of cations in the vacuolar lumen (Nhx1 and Vnx1), a third system, Vhc1, a cation-chloride co-transporter, has been identified in our studies (Petrezselyova, Kinclova-Zimmermannova, & Sychrova, 2013). Also mitochondria contain a highly efficient system that exchanges potassium cations for protons. However, this transport system, as shown within the Translucent project, is composed of several small proteins that together form an efficient transporter (Zotova et al., 2010).

2.3. Interactions of ion transporters with other proteins

Since the completion of the *S. cerevisiae* genome sequencing, numerous large-scale studies have been conducted to obtain an organism-wide overview of all protein–protein interactions. In addition, many protein–protein interactions have been found in studies addressing a single or a limited number of proteins. Techniques used to study protein–protein interactions include the yeast two-hybrid system, *in vitro* protein–protein interaction studies, co-purification and co-immunoprecipitation experiments. A major difficulty in analysing proteins interacting with ion transporters is the fact that ion transporters are membrane-bound, making them difficult to extract from yeast cell homogenates. Furthermore, for the classical yeast two-hybrid system, proteins are fused to parts of the Gal4 transcription factor. For the system to work, fusion proteins must be targeted to the nucleus, which may be difficult for membrane proteins. Despite these potential pitfalls, ion transporters were included in the large-scale studies and many physical interaction partners have been found (Fig. 1.2). Most of the interaction partners bind to only one of the transporters, suggesting that there is no common regulator for the various transporters. The interaction partners have many different cellular functions, most of them lacking an obvious link to ion homeostasis. Another important observation is that despite the fact that ion transporters are mainly found in the plasma and vacuolar membranes (see Section 2.1), many of the interaction partners are localized to different cellular organelles like the nucleus.

Because of the hydrophobic character of ion transporters, it is likely that many interaction partners have not yet been identified. To extract ion transporters from the membrane, strong detergents can be used. However, most of the interactions with other proteins will be lost. To study the interaction between Trk1 and the protein phosphatase Ppz1, cross-linking was used to maintain protein–protein interactions during extraction with strong

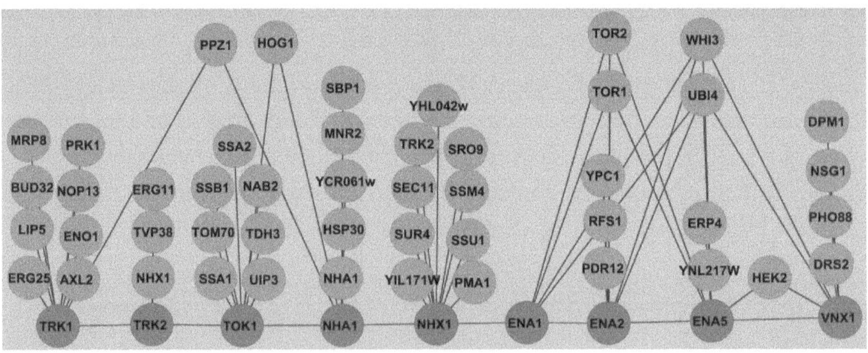

Figure 1.2 Overview of the proteins physically interacting with the major *S. cerevisiae* ion transporters. Data on protein–protein interactions were obtained from the Saccharomyces Genome Database (www.yeastgenome.org). The figure was prepared using Cytoscape (Cline et al., 2007).

detergents (Yenush, Merchan, Holmes, & Serrano, 2005). Such approaches may reveal a more complete picture of all interactions and potential regulators of the ion transporters.

2.4. Phosphorylation of ion transporters

The activity of ion transporters is mainly regulated at the post-transcriptional level. For example, ion transporters are phosphorylated at many sites. More specifically, large-scale studies on protein phosphorylation using mass spectroscopy revealed 15 phosphorylation sites in the Nha1 transporter, 13 in the Pma1 proton pump, 7 in the Trk1 transporter, and 1 in the Tok1 channel (PhosphoGrid, www.phosphogrid.org). The presence of so many phosphorylation sites suggests a very complex regulatory mechanism. The kinases involved in phosphorylation of these sites and the protein phosphatases involved in their dephosphorylation are largely unknown. Studies on individual transporters have identified a number of protein kinases and phosphatases. One of these is the Hog1 MAP kinase which may have an important function in the immediate response to high extracellular ion concentrations. Osmotic stress activates Hog1, which phosphorylates the Nha1 Na^+/H^+ antiporter and the Tok1 potassium channel (Proft & Struhl, 2004). Hog1 phosphorylation stimulates Nha1 activity, and this initial response to hyperosmolarity precedes and temporally regulates the activation of stress-response genes that depends on Hog1 phosphorylation of transcription factors in the nucleus. In this way, a very rapid stress relief is provided that facilitates subsequent changes in gene expression that permit long-term adaptation to harsh environmental conditions (Proft & Struhl, 2004). Other examples of protein kinases involved in the regulation of ion transporters are Sat4 and Hal5 (Mulet et al., 1999; Petrezselyova, Zahradka, & Sychrova, 2010). These homologous kinases are involved in the stabilization of the Trk1 transporter in the plasma membrane (Perez-Valle et al., 2007). They probably do not phosphorylate the transporter itself.

The Ppz1 phosphatase was shown to be involved in the regulation of the Trk1 potassium transporter (Yenush, Mulet, Arino, & Serrano, 2002) and it was demonstrated that Ppz1 co-immunoprecipitates with Trk1 (Yenush et al., 2005). In this same study, the *in vivo* phosphorylation of Trk1 was shown to increase in *ppz1 ppz2* mutants and this has been postulated to correspond to an increase in transporter activity.

14-3-3 Proteins bind to many phosphorylated proteins. In this way, they regulate the activity of the binding partners, their subcellular localization or their interaction with other proteins. Many plant and mammalian ion transporters are 14-3-3 binding partners. The yeast *S. cerevisiae* contains two 14-3-3 proteins, Bmh1 and Bmh2 (van Heusden, 2009). Recently, we have shown an interaction between Nha1 and the yeast 14-3-3 proteins (see also Section 2.3) (Zahradka, van Heusden, & Sychrova, 2012). However, until now the role of this interaction is unclear. 14-3-3 Proteins very likely interact with other transporters as well.

3. SETTING STRAINS, METHODS, REAGENTS AND STANDARDS

The groups involved in the Translucent project used various laboratory *S. cerevisiae* strains in their previous studies. As it is well known that many yeast cell features and phenotypes strongly depend on the genetic background, the consortium decided to work with only one strain. The selection of a proper strain was thoroughly considered and the BY4741 strain was ultimately chosen. Among the reasons for this choice are the availability of complete genome sequence, the existence of gene-deletion libraries, and the presence of suitable auxotrophic markers. The properties of the BY4741 strain were compared with those of the W303 strain (used in many previous studies of cation transport and homeostasis), and the results showed substantial differences (Petrezselyova et al., 2010), confirming the necessity to use the same set of strains in all partner laboratories and in all experiments.

Similarly, the planned work revealed the necessity to use the same growth media in all laboratories. Besides the standard YPD and YNB media, that usually contain approx. 15–20 mM K^+, the consortium needed a growth medium with very low-potassium content in order to study the effects of potassium starvation on cell physiology. For this purpose, a new synthetic minimal medium (K^+-free YNB-F without amino acids, ammonium sulphate and potassium; ForMedium™, UK) has been developed together with its low-fluorescence derivative (K^+-free YNB-F without riboflavin and folic acid; ForMedium™, UK) suitable for fluorescence measurements, for example, estimation of intracellular pH (cf. below). These media contain only 15 μM potassium, and thus can be brought to any desired potassium concentration by the simple addition of the appropriate salt. Secondly, when properly complemented, they sustain much more vigorous growth of yeast strains than K^+-free synthetic media previously reported (Rodriguez-Navarro & Ramos, 1984).

3.1. Construction of a series of mutants lacking potassium transporters

To characterize the function and activity of individual monovalent cation transporters, a series of 35 mutant strains derived from BY4741 was prepared using the Cre-*loxP* system (Supplementary Table 1, http://dx.doi.org/10.1016/B978-0-12-800143-1.00001-4). Deletion cassettes were designed to remove the entire open reading frame of the selected genes, using either the *KanMX* or *LEU2* marker genes or subsequent removal of the cassette by the Cre recombinase (Guldener, Heck, Fielder, Beinhauer, & Hegemann, 1996). In addition to the nine single mutants lacking individual cation transporters, double, triple, quadruple and quintuple mutants were prepared and characterized. In order to study the genetic interactions of genes encoding potassium transporters and genes encoding putative regulatory proteins, some other genes were deleted in strains lacking the transporters (e.g. *CNB1*, *CRZ1*, *HOG1*, *BMH1*, *HAL4* and *HAL5* genes; cf. e.g. Marquina et al., 2012; Perez-Valle et al., 2010; Zahradka & Sychrova, 2012). Altogether, the Translucent collection contains more than 60 strains.

3.2. Basic characterization of mutant strains

To characterize and compare the mutant strains, several approaches have been used. First of all, the growth phenotypes in both liquid and solid media supplemented with different amounts of KCl, adjusted to various pH, or containing cationic drugs were compared. Next, some crucial physiological parameters, such as the potassium content (using atomic absorption spectrophotometry, AAS), intracellular pH (with the use of expressed pHluorin), cell volume or relative membrane potential (3,3′-dipropylthiadicarbocyanine, diS-C_3(3) assay) were estimated, and finally the uptake or efflux of potassium (or its analogue rubidium) was measured under various conditions. As a standard, the parameters of the parental BY4741 strain were used. Figure 1.3A shows that the intracellular content of K^+ in BY4741 cells depended on extracellular KCl, mainly at low extracellular concentrations. This difference was also observable as a difference in the cell size, cells containing higher amount of potassium being bigger (not shown). Increasing the external potassium concentration diminished the relative membrane potential, but did not change the intracellular pH (Fig. 1.3B and C).

Characterization of mutants revealed many interesting phenotypes and changes in physiological parameters. Some examples are shown in

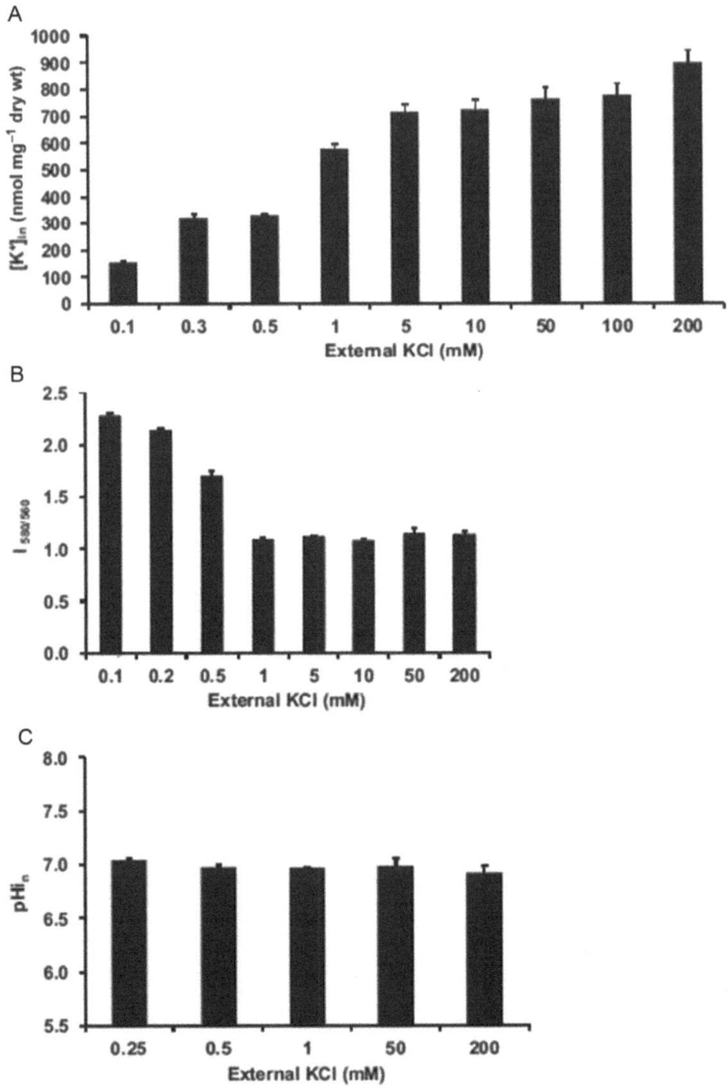

Figure 1.3 Basic physiological parameters of BY4741 grown in YNB-F supplemented with various potassium concentrations. Cells were grown in YNB-F supplemented as indicated to mid-logarithmic phase and the intracellular concentration of potassium (A), relative membrane potential (B) and intracellular pH (C) were estimated.

Fig. 1.4. As expected, deletion of genes encoding the potassium uptake systems abolished the ability to grow on low potassium, whereas the tolerance to high external KCl was diminished in strains lacking the potassium efflux systems (Fig. 1.4A). Surprisingly, the strains lacking the Trk1 and Trk2

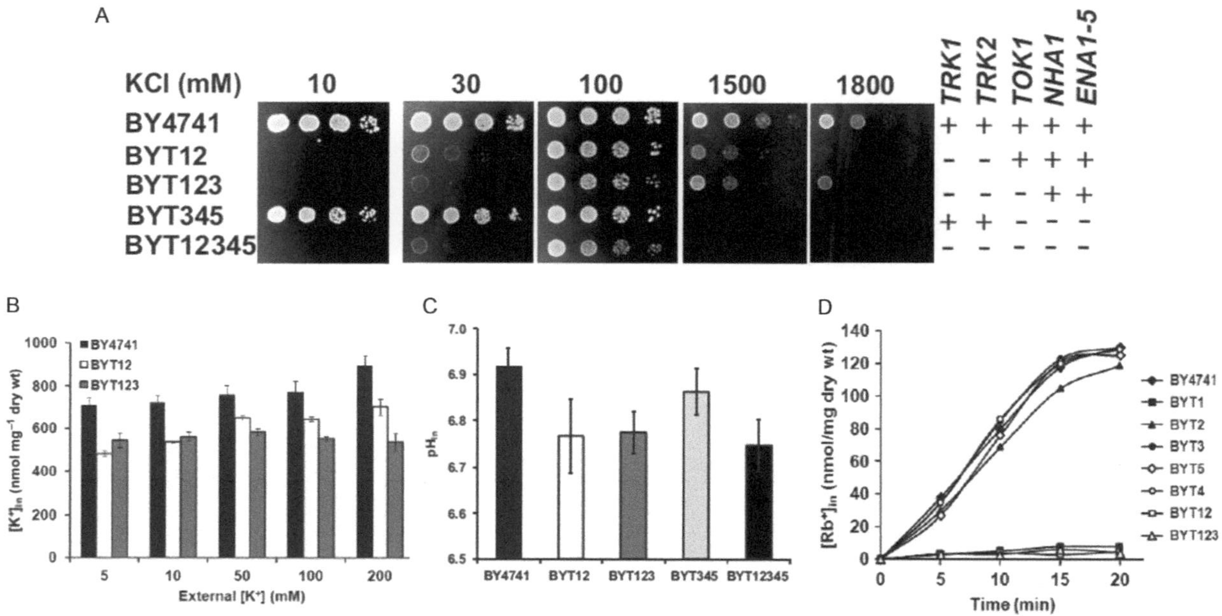

Figure 1.4 Comparison of growth phenotypes and physiological parameters of strains lacking potassium importing and/or exporting transport systems. (A) Growth of the parental strain (BY4741) and strains lacking either potassium uptake systems (BYT12) or potassium efflux systems (BYT345) or both (BYT123 and BYT12345) on YNB-F plates supplemented with KCl as indicated. (B) Intracellular potassium content in strains grown in YNB-F supplemented with different external KCl concentrations. (C) Intracellular pH in cells grown in YNB-F supplemented with 50 mM KCl. (D) Uptake of 0.5 mM Rb^+ into cells grown in YNB-F supplemented with 50 mM KCl and starved in the same medium without potassium during 2 h.

uptake systems also grew worse than the parental BY4741 in media with very high KCl (Fig. 1.4A). Measurements of membrane potential showed that this was due to the hyperpolarization of the plasma membrane of BYT12 (*trk1 trk2*) cells (Navarrete et al., 2010; Petrezselyova et al., 2011), as previously shown in a different genetic background (Madrid, Gomez, Ramos, & Rodriguez-Navarro, 1998), that resulted simultaneously in higher non-specific influx of cations and diminished activity of potassium exporters (Zahradka & Sychrova, 2012). Figure 1.4B shows that the deletion of Trk uptake systems resulted in a lower intracellular potassium content compared to cells with functional importers, which was accompanied by a lower intracellular pH (Fig. 1.4C). Figure 1.4D shows that Trk1 is the major system for the uptake of the potassium analogue rubidium and that deletion of the corresponding genes resulted in the loss of transport activity (strains BYT1, BYT12, BYT12345).

Altogether, a detailed characterization of mutants grown in standard conditions and in extremely low/high external concentrations of potassium revealed many new findings and simultaneously generated data that was used for modelling purposes (cf. Section 7). Our results showed that the loss of potassium uptake activity is compensated by a decreased efflux of potassium, that the potassium fluxes across the plasma membrane influence the activity of the Pma1 H^+-ATPase and thereby the intracellular pH, that the vacuolar potassium transporters are involved in adaptation to osmotic stress, etc. (Barreto et al., 2011; Kahm et al., 2012; Navarrete et al., 2010; Petrezselyova et al., 2013; Zahradka & Sychrova, 2012).They also brought about the necessity to analyse the intracellular distribution of potassium among the organelles in more detail.

3.3. Intracellular potassium distribution

The only relatively widely used approach to investigate intracellular distribution of potassium among the organelles is the use of substances such as cytochrome *c*, DEAE dextran or digitonin to permeabilize the plasma membrane. The main limitation of these procedures is that the conclusions are too simplistic, since they usually consider only two important compartments where the cations are present in significant amounts; potassium extracted after permeabilization is considered as the cytoplasmic fraction, whereas the remaining amount is ascribed to the vacuole. After using these procedures, most of the investigations conclude that cations accumulate in the vacuolar fraction to significantly higher amounts than in the cytoplasmic fraction (Okorokov, Lichko, & Kulaev, 1980). This affirmation contradicts the only previously published results, using energy-dispersive X-ray microanalysis, from which it was concluded that the intracellular distribution of potassium in the cytoplasm and in the nucleus was about equal and twice the amount present in the vacuole (Roomans & Seveus, 1976). By preparing spheroplasts, subsequent homogenization and differential centrifugation, we were able to isolate highly pure fractions corresponding to vacuoles, mitochondria, endoplasmic reticulum (ER), Golgi apparatus, nuclei and cytosol, and we determined their potassium content by AAS. We found that while vacuoles accumulated most of the intracellular potassium, the cytosol contained relatively low amounts of this cation. However, the concentrations of potassium in the cytosol were kept rather constant during the potassium starvation process and for that purpose vacuolar potassium had to be rapidly mobilized. We also found that the nuclei had and maintained a relatively constant and important percentage of the total intracellular potassium. Finally, minor amounts of potassium were measured in ER, the GA and mitochondria (Herrera, Alvarez, Gelis, & Ramos, 2013). Table 1.2 shows, as an example, the intracellular distribution of potassium in BY4741 cells grown in YNB-F with 50 mM KCl.

3.4. FLISE as a simple method to analyse net ion fluxes across the plasma membrane

3.4.1 Introductory concepts

The knowledge about ion fluxes and the membrane potential across the plasma membrane are important prerequisites for the analysis of ion homeostasis. The standard way to obtain such information is the application of electrophysiological techniques like patch or voltage clamp. However, due to the presence of a cell wall, prohibiting the formation of a "giga seal" and that must be enzymatically removed, the patch clamp technique cannot be easily applied to yeast cells (see also Section 3.5.1). The enzymatic removal of the cell wall also might influence the properties of the proteins involved in mediating the currents and distort the results. Thus, only relatively few electrophysiological studies on yeast plasma membrane exist to date (Bertl, Bihler, Kettner, & Slayman, 1998; Bertl, Bihler, Reid, et al., 1998; Bertl et al., 2003; Haro & Rodriguez-Navarro, 2002, 2003; Rivetta, Kuroda, & Slayman, 2011). Within Translucent several approaches were used to overcome these difficulties: The fluorescent dye diS-$C_3(3)$ was used to monitor qualitative differences in membrane potential (Petrezselyova et al., 2011; Zahradka & Sychrova, 2012) and to develop a

Table 1.2 Subcellular potassium distribution in BY4741 cells grown in YNB-F supplemented with 50 mM KCl

	ER	GA	Mit	Nucleus	Vacuole	Cytosol
K^+ (nmol mg^{-1})	27.9 ± 3.1	10.8 ± 2.3	33.9 ± 3.3	164.1 ± 12.6	253.5 ± 11.4	73.7 ± 6.5
%	4.95	1.92	6.01	29.10	44.96	13.06

Data are mean ± SEM from at least four independent experiments with triplicate determinations (two experiments for ER and GA data): ER, endoplasmic reticulum; GA, Golgi apparatus; Mit, mitochondria.

method to quantify changes of membrane potential in absolute (mV) values (Plášek, Gášková, Lichtenberg-Fraté, Ludwig, & Höfer, 2012). This method was used for assessment of plasma membrane depolarization in *S. cerevisiae* induced by transmembrane ion fluxes (Plášek, Gášková, Ludwig, & Höfer, 2013). To assess ion fluxes directly, we developed a system that can be easily used to monitor these fluxes across the plasma membrane of yeast (or other) cells in suspension.

Flux measurements using ion-selective electrodes (FLISE) is based on the highly time-resolved monitoring of the concentration of the analysed ion ($[ion]_{ext}$) in cell suspension with ion-selective electrodes (ISEs). A scheme of the FLISE set-up is shown in Fig. 1.5A. The voltage data, that is, the potential difference between an ISE and a reference electrode, are recorded on a PC via high impedance amplifiers and A/D converters. For data analysis (Fig. 1.5B), a smoothing spline is applied. Voltages are converted to $[ion]_{ext}$ according the relation $[ion]_{ext} = 10^{(V\text{-intercept})/\text{slope}}$. The flux of the respective ion (i.e. the change of its concentration vs. time) is then calculated as the numerical derivative $d[ion]_{ext}/dt$ of the $[ion]_{ext}$ (time) function, resulting in the flux from the suspension medium into the cells or vice versa. The ion fluxes across the plasma membrane are then assessed by determining optical density of the experimental cell suspension and by its conversion to cell number and membrane area. Measurements can be performed in total volumes of less than 2 mL. Two examples for the application of FLISE within Translucent are given below.

3.4.2 Simultaneous measurements of K^+ uptake and H^+ extrusion in starved cells and upon energization

For the experiments, BY4741 cells were grown to stationary phase and starved in distilled water over night at 4 °C. The next day the cells were collected by centrifugation and resuspended in 1 mM MES/NaOH pH 5.0, 10 mM $CaCl_2$. Suitable aliquots were then transferred to the FLISE cuvette. KCl was added to the desired concentration and cells were energized by the addition of glucose through activation of the plasma membrane proton pump Pma1 (Eraso, Mazon, & Portillo, 2006; Eraso & Portillo, 1994; Serrano, 1983) and stimulation of glycolysis. In Fig. 1.6A, the results of a representative experiment are shown. Approximately 1 min after adding glucose, H^+ efflux and K^+ influx started fairly simultaneously and increased to a transient peak. Subsequently, the fluxes declined again. In contrast to K^+ influx that vanished completely (and often even reverted to an efflux at the end of the experiment), a small steady-state H^+ efflux persisted. Not only K^+

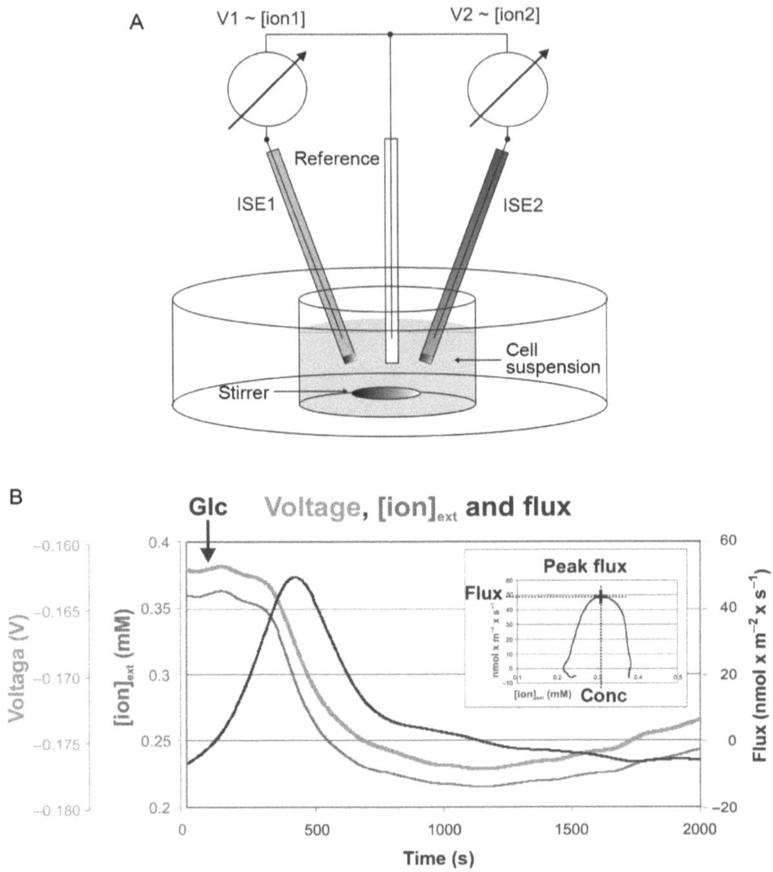

Figure 1.5 The principle of FLISE. (A) Scheme of the FLISE set-up. The potential differences (voltages, V) between ion-selective electrodes (ISEs) versus a reference electrode are measured in a cuvette containing a suspension of cells. Data are transferred to a PC via high impedance amplifiers and A/D converters (not shown). Cells are kept in suspension by stirring. Temperature is kept constant. (B) Conversion of measured voltages to ion flux. Recorded voltages are after noise reduction *b* median filter and spline fit transformed to external ion concentrations ($[\text{ion}]_{\text{ext}}$) using the function $[\text{ion}]=10^{(\text{V-intercept})/\text{slope}}$. Slope and intercept are determined for each experiment by calibration with different $[\text{ion}]_{\text{ext}}$ by linear regression fit of $V=\text{slope}^*\log([\text{ion}]_{\text{ext}})+\text{intercept}$. The flux that is given by $\text{flux}=d([\text{ion}]_{\text{ext}})/dt$ ($t=\text{time}$) is obtained by numerical derivation of the $[\text{ion}]_{\text{ext}}(t)$ function and is related to the membrane area using the known cell number and average size of cells. Because FLISE is based on the measurement of external concentration changes, fluxes do not correspond to the initial but to the actual $[\text{ion}]_{\text{ext}}$. The external concentration at which peak fluxes occur is thus determined by plotting flux versus $[\text{ion}]_{\text{ext}}$.

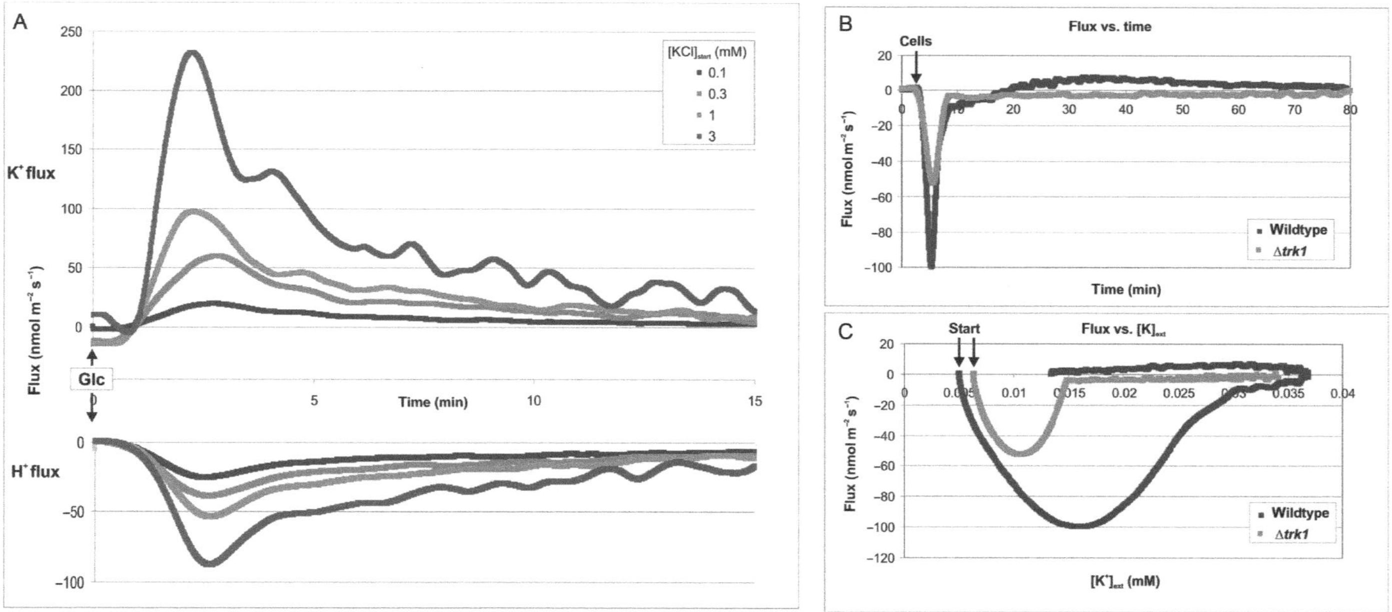

Figure 1.6 Ion fluxes determined using FLISE. (A) Simultaneous recorded K^+ and H^+ fluxes at different added $[KCl]_{start}$ as indicated across the plasma membrane of BY4741 wild-type cells after starvation. Cells were energized by addition of glucose at time 0. Not only potassium uptake (positive fluxes) but also H^+ extrusion (negative fluxes) were dependent on $[K^+]_{ext}$. However, the fluxes do not sum up to 0. Whereas at low $[K]_{ext}$ proton efflux was larger than K^+ uptake, the ratio reverted at higher external K^+ concentrations. The apparent "oscillations" observed with 3 mM $[K^+]_{ext}$ are artefacts caused by smoothing and numerical calculation of the derivative. (B) Time course of K^+ efflux during potassium starvation. Cells were added to nominally K^+-free medium at the time indicated and flux was monitored for 80 min. Wild-type cells showed a transient efflux that eventually reverted to K^+ re-uptake. *trk1*Δ cells exhibited less efflux that continued during the course of the experiment. (C) Plot of K^+ fluxes shown in *B* versus $[K^+]_{ext}$. The external K^+ concentration had only increased to ~37 μM at the onset of re-uptake by wt cells and decreased to less than 15 μM. (See the color plate.)

influx but also H^+ efflux increased with $[K^+]_{ext}$ indicating that H^+ extrusion and K^+ uptake are coupled through charge compensation. However, the amounts of translocated K^+ and H^+ were approximately equal only in the range of ~0.3–1 mM $[K^+]_{ext}$. At lower $[K^+]_{ext}$, more protons were extruded than K^+ taken up and at higher $[K^+]_{ext}$ more K^+ was taken up than protons extruded. The apparently lower H^+ extrusion at high $[K^+]_{ext}$ could be due to an error in H^+ flux estimation: bicarbonate formed in the process of glycolysis is exported and thus may buffer $[H^+]$ changes in the medium. This would lead to a decrease of $[H^+]_{ext}$ and thus to an underestimation of H^+ efflux. However, there is no easy explanation for the apparent excess of proton efflux at low $[K^+]_{ext}$. Thus, it seems likely that other yet unspecified ions are translocated across the plasma membrane in parallel.

3.4.3 K^+ efflux and re-uptake upon potassium starvation in BY4741 wild type and trk1△ cells

One of the goals of the Translucent consortium was the analysis and modelling of K^+ homeostasis during K^+ starvation (Kahm et al., 2012; Navarrete et al., 2010). In these experiments, cells grown in non-limiting $[K^+]$ (50 mM) were harvested in exponential phase and transferred to K^+-free medium. The K^+ flux was subsequently calculated from the internal K^+ concentrations determined by atomic absorption spectroscopy at various times during several hours. The necessary washing steps led to a delay in which cells could theoretically loose K^+ unmonitored and the method does not allow good time resolution due to the steps involved. In order to (i) decrease the time gap between removal of $[K^+]_{ext}$ and the first measurement and (ii) to monitor K^+ flux in high time resolution, we performed FLISE experiments. Cells were harvested by centrifugation, washed once with K^+ free medium, resuspended and immediately transferred to the measurement cuvette. By this short protocol, we could reduce the time span of unmonitored K^+ loss to less than 2 min. For the experiments, we used cells in the exponential growth phase.

It must be mentioned that these growth and analysis conditions slightly differed from those used by Navarrete et al. (2010). YNBF (the Translucent medium) contains as nitrogen source NH_4^+, an ion that is known to compete for K^+ in many K^+ translocation systems, both natural and artificial (including valinomycin, the K^+-ionophore used in K^+-ISEs). Hence, we used SDAP, a medium with arginine as the nitrogen source (Rodriguez-Navarro & Ramos, 1984). Because we were interested in K^+ fluxes only, $[H^+]_{out}$ was kept constant by buffering.

The results of representative starvation experiments with BY4741 (wild type) cells and the corresponding *trk1Δ* strain under these conditions are shown in Fig. 1.6B and C. With BY4741 (wild type) cells, a fast transient K^+ efflux with a peak approximately 2 min after cell transfer to medium without added KCl was observed (Fig. 1.6B). Following a sharp decrease of K^+ efflux, a short plateau lasting a few minutes was reached continuing with further decrease of efflux that after ~20 min reverted to influx. In contrast, with *trk1Δ* cells re-uptake of K^+ was never observed. In addition, the initial K^+ efflux was lower in *trk1Δ* cells as compared to the wild-type strain.

A plot of flux versus $[K^+]_{ext}$ from the same experiment (Fig. 1.6C) revealed that the re-uptake was not caused by a strong increase in $[K^+]_{ext}$ that could cause a reversal of the electrochemical K^+ gradient. The onset of re-uptake occurred at low $[K^+]_{ext}$ (less than 40 μM $[K^+]_{ext}$). Subsequently, $[K^+]_{ext}$ decreased to less than 15 μM. In some cases, re-uptake even led to final $[K^+]_{ext}$ values that were lower than that at the beginning of the experiment (before addition of cells). Thus, it appears likely that during the experiment a transition from a medium to a high-affinity mode of K^+ uptake took place. Such a difference in the kinetics of K^+ uptake was already described by Rodriguez-Navarro and Ramos (1984). They compared K^+ uptake between cells grown in high and low $[K^+]_{ext}$ and found that cells grown in low $[K^+]_{ext}$ possessed higher affinity K^+ uptake system(s) compared to cells grown in non-limiting $[K^+]_{ext}$. This finding was later attributed to a switch between medium and high Trk1 K^+ affinity modes (Haro & Rodriguez-Navarro, 2002), with the high affinity being possibly caused by phosphorylation of Trk1 (Yenush et al., 2005).

3.4.4 FLISElab: An integrated tool for evaluating and documenting FLISE data

FLISE raw data consist of series of time and voltage values that have to be annotated, denoised/smoothed, converted to [ion] and translated to flux in a reproducible, easy to handle way. To reduce the gap between experiment and final data storage, a time-series pre-processing tool, FLISElab, was developed. FLISElab allows data description and the addition of unrecorded inputs (e.g. time and nature of injections like KCl or glucose; cf. Fig. 1.6A). As multiple experiments are usually carried out on 1 day and stored in the same (raw-) data file, the tool segments the raw data files into single-experiment files together with all the necessary information to (pre-)process them. A module for smoothing and differentiation is included in FLISElab. The resulting files that are human (data sheet) and machine-readable can be transferred to the Translucent data management system, pyMantis. FLISElab is available for Windows, Linux and Os X (http://translucent-network.org/FLISELAB/plugin_release/download) and can also be used online. During the testing period with continuous discussion between developers and experimentalists, FLISElab became a robust tool, adapted to the needs of users.

In conclusion, within the Translucent project we developed FLISE, a simple and reliable method to analyse ion fluxes across the plasma membrane of (yeast) cells in suspension. Due to the logarithmic correlation between voltage and ion concentration changes, FLISE is most useful in the low ion concentration range. By using high cell densities, it is possible to calculate fluxes in suspensions containing $[\text{ion}]_{\text{ext}}$ of ~10 mM. In contrast to radioactive tracer methods, FLISE can only be used to determine the total net fluxes of ions across the plasma membrane and is not suitable for analyses of unidirectional fluxes. In contrast to classical electrophysiology, however, FLISE is also suited to analyse electroneutral ion transport processes, for example, the antiport of cations having a stoichiometry of 1/1.

3.5. Electrophysiology of yeast cation transporters

3.5.1 Preparation of yeast membranes for electrophysiological characterization

A study of yeast ion channels would be incomplete without the use of electrophysiology to study the unitary conductances found in yeast membranes. Unfortunately, as mentioned in Section 3.4.1, yeast membranes are not very amenable to patch clamp recording, due mainly to the presence of a cell wall and their small cell size of approximately 5 μm. Complete removal of the cell wall does not generally produce reliable membranes for patch clamping due principally to the amount of cell wall debris produced by this process which remains attached to the plasma membrane. The most successful method utilized to patch clamp the yeast plasma membrane involves partial removal of the cell wall followed by induction of membrane blebs through gaps in the cell wall by alteration of the osmotic strength of the external buffer (Bertl, Bihler, Kettner, et al., 1998). However, this method produces unreliable patches which are often unstable, an important problem when attempting experiments requiring multiple external buffer changes. Other researchers have utilized a giant cell method in order to increase the chances of achieving reliable seals (Bertl, Bihler, Kettner, et al., 1998). Many factors can affect the chances of achieving reliable patch pipette seals. The choice of growth media can alter the yeast cell wall properties, which appears to affect the chances of obtaining clean plasma cell membranes greatly. A side-by-side electrophysiological comparison of yeast strains containing a gene deletion of *NHA1* replaced with either a cassette providing neomycin tolerance or the *LEU2* gene and grown in either YPD- or SD-based media as appropriate was undertaken. Whole-cell gigaohm seals were obtained in 9/100 and 2/100 of attempts, respectively (E. Aydar & C. P. Palmer, Translucent-2, unpublished data).

Compared to the yeast plasma membrane, clean yeast vacuolar membranes can be easily prepared and assessed with the patch clamp pipette (Palmer et al., 2001). Yeast spheroplasts can be prepared and subsequently the plasma membrane can be easily removed by rupturing using a buffer of lower osmotic strength and continuous perfusion to remove cell debris. This method has been utilized very successfully to observe the vacuolar current Yvc1 (Palmer et al., 2001) (aka TrpY1) or to heterologously express vacuolar channels from other yeast species (Zhou, Loukin, Coria, Kung, & Saimi, 2005). Using this method, a variety of patch pipette configurations can be achieved including whole-vacuole, inside out and outside in detached patches allowing manipulation of both extravacuolar and intravacuolar ionic concentrations.

3.5.2 Utilization of automated patch clamp systems and electrofusion

With the advent of automated patch clamp systems which are more financially attainable, researchers have begun investigating their potential use for surveying yeast membranes. The advantage being that a number of yeast can be applied to the system at one time thus increasing the chances of achieving a successful and stable patch pipette seal. As such, only strongly sealed spheroplasts need to be further assessed, thus decreasing the time taken to obtain reliable data. Additionally, the format of the automated patch allows the spheroplasts to be sucked onto the recording apertures. Recently, researchers have utilized electrofusion of spheroplasts in order to obtain giant spheroplasts and increase the chances of obtaining reliable seals (Terpitz et al., 2008).

3.5.3 Summary of the electrophysiological characterization of yeast cationic channels to date

To date, only a few yeast cation conductances have been characterized using the patch pipette, including Tok1 (Bertl, Bihler, Reid, et al., 1998), Trpy1 (Bertl & Slayman, 1990), Mid1 (Kanzaki et al., 1999), Trk1 and Trk2 (Bertl et al., 2003). Notable exceptions include Nha1, Cch1 and Ena1, for which no electrophysiological recordings have to date been reported. It has been suggested that the Nha1 transporter is electrogenic (Ohgaki, Nakamura, Mitsui, & Kanazawa, 2005); however, transient expression of Nha1 in HEK-293 cells using expression vector pCI-neo does not give rise to any new conductance when compared to mock-transfected cells (E. Aydar & C. P. Palmer, Translucent-2, unpublished data). In contrast, a number of cationic conductances have been recorded from yeast plasma membranes for which no known gene has as yet been associated with, including a prominent mechano-sensitive ion channel (Gustin, Zhou, Martinac, & Kung, 1988). It should be noted, however, that knock out of the *MID1* gene does not remove the mechano-sensitive channel current which can be recorded from the plasma membrane (C. Kung & C.P. Palmer, unpublished communication). Additionally, a high non-selective cation current was previously described by whole-cell patch clamp of yeast plasma membranes, but to date the gene encoding this current has yet to be identified (Bihler, Slayman, & Bertl, 1998).

3.5.4 Expression of yeast ion channels and transporters in eukaryotic expression systems

Heterologous expression systems have been utilized frequently to express mammalian ion channels and transporters. However, to date this technique has not been widely used to study yeast channels (Table 1.3). The advantages of these systems include the capacity to over express the channel/transporter, which increases the chances of obtaining reliable data, the ease with which cell membranes can be approached with a patch pipette and the stability of the gigaohm seals produced.

All these factors are influenced by the lack of debris which is associated with the membranes in these systems compared to the yeast plasma membrane which invariably has partially digested cell wall components attached to it that can influence patch pipette sealing and the quality of seals achieved. Oocytes prepared from *Xenopus laevis* have been successfully utilized to express and characterize currents expressed by the *TOK1* gene (Loukin et al., 1997). The *MID1* gene can be expressed in mammalian expression systems, such as CHO cells and a stretch sensitive cation channel was recorded (Kanzaki et al., 1999). However, as mentioned above, gene deletion of *MID1* in yeast does not remove the prominent mechano-sensitive cation channel electrophysiologically recorded from yeast plasma membranes (Gustin et al., 1988). The *NHA1* gene was successfully transiently expressed in HEK cells as visualized by green fluorescent protein (GFP)

Table 1.3 Heterologous expression systems utilized to express *S. cerevisiae* cation channels and transporters

Gene	Expression system	Successful expression	Electrophysiological recordings	References
TOK1	*Xenopus laevis* oocytes	Yes	Yes	Loukin et al. (1997)
YVC1	*Xenopus laevis* oocytes	No	–	Kung and Palmer (unpublished communication)
TRK1	Transient plasmid expression Vector: pCI-neo HEK cells	No	–	Translucent unpublished data
NHA1	*Xenopus laevis* oocytes Transient plasmid expression Vector: pCI-neo COS-7	Yes Yes	No: masked by an endogenous current No	Translucent unpublished data
MID1	COS-7 CHO	Yes	Yes	Kanzaki et al. (1999)

expression with a tagged construct using the CMV-based plasmid pCI-neo. However, to date no additional currents have been recorded in cells expressing this plasmid indicating that either the transporter is not electrogenic or the transporter is a low volume transporter, a common problem when studying transporters (E. Aydar & C. P. Palmer, Translucent-2, unpublished data).

4. POTASSIUM HOMEOSTASIS IN *S. CEREVISIAE*: AN OMICS APPROACH

4.1. Identification of mutants with altered potassium homeostasis

Potassium uptake is the principal consumer of the electrochemical gradient generated by the Pma1 proton ATPase (Goossens, de La Fuente, Forment, Serrano, & Portillo, 2000; Madrid et al., 1998). We considered that mutations simultaneously causing altered tolerance to diverse toxic cations, differing in their specific transport systems and toxicity mechanism(s), would probably reflect changes in the electrochemical gradient and, consequently, would lead to the identification of relevant components in this fundamental process. The screen of the *S. cerevisiae* haploid deletion library (Barreto et al., 2011) for changes in tolerance to three toxic cations, hygromycin B, spermine and tetramethylammonium, yielded 226 mutant strains displaying altered tolerance to all three drugs (202 sensitive and 24 tolerant). From this set, 123 strains showed a strong (60 strains) or moderate (63 mutants) growth defect in the presence of limiting concentrations of potassium (1 mM). Hence, in accordance with our expectations, a significant number of the selected strains (54%) had altered potassium requirements. A Gene Ontology analysis of the mutants revealed significant enrichment in several cellular processes, such as post-translational processing (mostly cells lacking genes encoding protein kinases and protein phosphatases), intracellular trafficking (Golgi compartment, endosome, and vacuolar transport), chromatin modification, and the regulation of transcription.

In many cases, our screen revealed highly coherent phenotypes for different mutations affecting a given function (Fig. 1.7A). A representative example is that, among the mutations conferring drug sensitivity, we have identified not only *YPT6*, which encodes a GTPase involved in the secretory pathway required for the fusion of endosome-derived vesicles with the late Golgi compartment, and which have been shown to participate in salt tolerance (Maresova, Vydareny, & Sychrova, 2012), but also *RGP1* and *RIC1*, encoding the exchange factor that catalyses nucleotide exchange on Ypt6; *SYS1*, which encodes a multicopy suppressor of the *ypt6* null mutation; and even *VPS63*, a dubious open reading frame, which overlaps with *YPT6* (and therefore, should generate similar phenotypes when deleted).

We then separated the mutants in four groups (A, B, C and D), on the basis of their tolerance to the drugs and their requirements for potassium. Thus, group A represented sensitive mutants with growth defects in low-potassium medium. Group B mutants were sensitive to the three tested cations, but grew nearly as well as the wild-type control in low-potassium medium. The tolerant mutants were similarly divided into group C (poor growth in low-potassium medium) and group D (no defect in medium with a low-potassium concentration). Then, a subset of 27 mutants representative of each group were selected and subjected to diverse biochemical tests, such as membrane potential, rubidium (a potassium analogue) uptake, presence of Trk1 in plasma membrane fractions, proton pumping (measured as the acidification of the medium and the presence of Pma1 in plasma membrane fractions) and *ENA1* ATPase expression. For a number of cases, it was possible

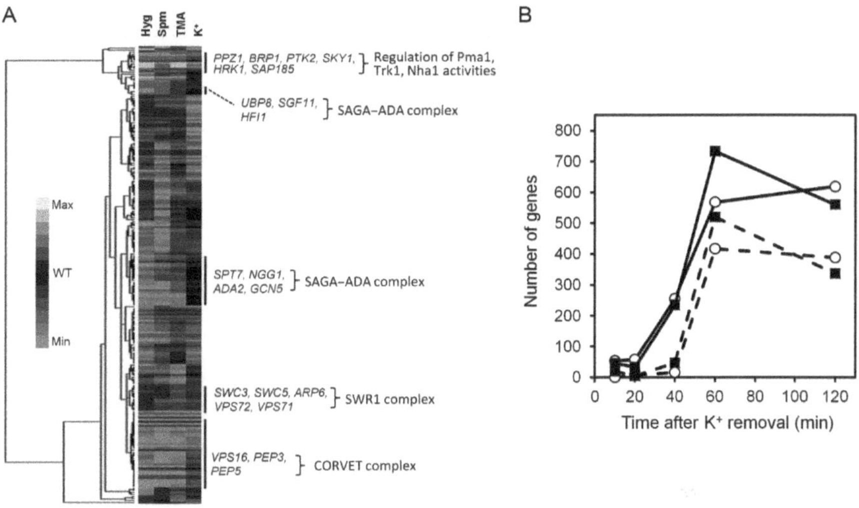

Figure 1.7 The OMICS of potassium homeostasis. (A) Heat-map displaying clustering analysis (Cluster v3.0, settings, correlation uncentered and average linkage) of tolerance to drugs and potassium requirements for the 226 mutants selected in the screen reported by Barreto et al. (2011). Phenotypes similar to the wild-type (WT) strain are shown in black. Hyg, hygromycin B; Spm, spermine; TMA, tetramethylammonium; K^+, growth at limiting potassium (1 mM). See the above for details. (B) The short-term transcriptional response to potassium starvation. The number of genes induced (circles, continuous line) or repressed (squares, continuous line) at a given time after removal of potassium from the medium is plotted. Discontinuous lines denote the number of genes that shows the peak of induction or repression at a given time. (See the color plate.)

to provide an explanation for the observed phenotypes. Thus, most mutants belonging to group A may suffer a defect in Trk1 function, whereas the phenotypes of group B strains would be independent of Trk1 and Pma1. In contrast, decreased levels of Pma1 expression or activity (which could indirectly affect Trk1 function) may explain the behaviour of group C mutants. The molecular bases for group D phenotypes were not obvious, although it may involve alterations of Nha1 activity, at least in some cases.

4.2. Transcriptomic profiling of potassium starvation provides clues on diverse physiological roles for potassium

The maintenance of high concentrations of intracellular potassium has been associated to diverse cellular functions in yeast, such as the regulation of cell volume and intracellular pH (Merchan, Bernal, Serrano, & Yenush, 2004), the maintenance of stable plasma membrane potential and compensation of negative charges in many macromolecules. Potassium has been reported for many years to be required for protein synthesis (Lubin & Ennis, 1964) and for the activation of diverse enzymes, such as pyruvate kinase (see Page & Di Cera, 2006). However, an overall characterization of the cellular roles for potassium was still pending and it was generally accepted that additional, still uncharacterized, cellular targets may exist. The effects of potassium limitation by transcriptomic profiling in yeast steady-state cultures growing under chemostat conditions was investigated a few years ago (Hess, Lu, Rabinowitz, & Botstein, 2006). Although this study revealed that when external potassium is low extracellular ammonium becomes toxic and impairs growth, only a small number of genes were identified as transcriptionally affected by potassium limitation. We considered that this limited response was not in agreement with the key role of potassium in the yeast physiology and, consequently, we developed a complementary approach based in the use of the Translucent K^+-free medium and aiming to investigate the short-term effects on gene expression profile derived from the sudden depletion of potassium in the environment.

We demonstrated that transfer of wild-type *S. cerevisiae* cells to K^+-free medium results in a strong short-term transcriptional response that affects over one thousand genes. The highest number of induced or repressed genes was observed after 60–120 of potassium starvation. Most induced genes showed a peak of expression at 60 min (Fig. 1.7B). Gene Ontology analysis of the genes induced by lack of potassium revealed a potent induction, even at short times, of the entire pathway for sulphur assimilation and methionine uptake and biosynthesis. This was confirmed for several genes (*CYS4*, *MET6*, *MET10*, *MUP3*, *SAM3*, *SER3*, *STR3* and *SUL3*) by the use of LacZ fusions of the promoters. These changes likely reflect potassium starvation-driven perturbation of sulphur-related metabolism, since we also observed that the intracellular levels of methionine and cysteine were almost immediately depleted in cells shifted to K^+-free medium, and that the induction of these genes is specifically attenuated by increasing the amount of methionine in the medium. Further evidence for the impact of alteration of normal potassium metabolism on this pathway is provided by the observation that genes belonging to the sulphate assimilation and methionine biosynthesis pathways are specifically induced in *hal4 hal5* mutants growing in standard YPD medium (Perez-Valle et al., 2010). Hal4 (Sat4) and Hal5 are protein kinases required for normal potassium uptake. A strain lacking both proteins fails to correctly allocate the potassium Trk1 transporter to the cell membrane under low-potassium conditions (Perez-Valle et al., 2007). In fact, the expression profile in *hal4 hal5* cells largely overlaps (correlation coefficient=0.77) with that determined for the *trk1 trk2* strain (Barreto, Canadell, Valverde-Saubi, Casamayor, & Arino, 2012).

Potassium starvation also triggered induction of *PHO*-regulated genes (i.e. *PHO84, PHO11, PHO12* and *PHO8*), as well as a number of genes essential to deal with oxidative stress and/or known to be induced under this condition (i.e. *CTT1, TSA2, PRX1, TRX3* and *SOD2*). The possibility that potassium starvation may trigger a situation of oxidative stress was confirmed by time-dependent monitoring of reactive oxygen species (ROS) formation with 1,2,3-dihydrorhodamine. Further confirmation came from the observation of a decline in reduced glutathione (GSH) levels and the accumulation of oxidized glutathione (GSSG), thus leading to a strong decrease in the GSH/GSSG ratio. Although at this point it is not obvious why potassium starvation would result in oxidative stress, it has been postulated that certain conditions leading to programmed cell death ROS production is inhibited by K^+ uptake (Hoeberichts et al., 2010). On the other hand, we also detected increased expression of *GLO1, GLO2* and *GRE3*, encoding enzymes required for detoxification of methylglyoxal, a by-product of glycolysis formed from triose phosphates. Indeed, a three- to fourfold increase in methylglyoxal levels was observed after 20 min of starvation. It has been postulated that an excess of methylglyoxal formation can cause oxidative stress (see Desai et al., 2010, for a review), and it is known that accumulation of this compound activates oxidative stress-responsive transcription factors in budding yeast (Maeta, Izawa, Okazaki, Kuge, & Inoue, 2004). Therefore, the early accumulation of methylglyoxal may explain, at least in part, the oxidative stress response observed under potassium limitation. Production of methylglyoxal and induction of trehalose metabolism have been linked in the past (Aguilera & Prieto, 2004). Interestingly, we also observed a strong increase in expression of genes related to trehalose metabolism (both synthesis and degradation) and we determined a peak of trehalose content shortly after potassium removal, followed by a rapid decrease in the levels of the disaccharide. This suggests that accumulation of methylglyoxal would trigger a futile cycle based on trehalose generation and degradation, which would help to down-regulate carbon flux through the upper part of the glycolysis. Such a process has been suggested to occur in response to other stresses (Parrou, Teste, & Francois, 1997). The overview of the transcriptomic response indicates that a substantial number of stress-responsive genes are also induced by potassium starvation. It must be noted, however, that lack of potassium in the medium did not provoke the activation of an integrated stress-response element

(STRE)-driven LacZ reporter, which is highly sensitive to the activation of the general stress response (Estruch, 2000). Therefore, the strong and diverse transcriptional response in cell exposed to potassium starvation is not mediated by activation of the general stress-response mechanisms.

Wild-type *S. cerevisiae* can grow at appreciable rate even when the K^+ concentration in the medium is rather low, but depletion of potassium ions below a certain threshold (in the low micromolar range for the wild type) leads to substantial growth inhibition (Navarrete et al., 2010). Evaluation of the transcriptional profile in response to the lack of potassium provides several clues to explain this growth arrest. Firstly, among the genes whose mRNA levels are depleted, there is a vast excess of those related to ribosome construction: genes encoding ribosomal proteins, rRNA processing genes and members of the so-called RIBI regulon, including genes important for the biosynthesis of GMP from IMP. This transcriptional response fits with the blockage in protein synthesis previously reported (Lubin & Ennis, 1964) and confirmed in our proteomic analyses (see Section 4.3). In addition, the level of diverse G1- and G2-phase cyclins mRNAs drastically decline upon K^+ starvation (mainly *CLN2*, *CLN3*, *CLB1*, *CLB2* and *CLB6*), and this is reflected in very low levels of the encoded proteins after 1–2 h upon starvation. Finally, septins are evolutionarily conserved filament-forming proteins whose localization at the bud neck is required for proper shaping of the bud. We observed that K^+-starved cells are unable to construct proper septin rings. It is unlikely that this defect is caused by degradation of septins, since at least Cdc11 levels remains stable in potassium-starved cells long after septin rings are disassembled. Since there is evidence that diverse cyclins, particularly Cln1 and Cln2, are needed to promote proper septin ring formation, the depletion of various cyclins as a result of potassium starvation might be the cause of the failure to assemble the septin ring.

4.3. Two-DE-based proteomics analysis of *S. cerevisiae* wild type and K^+ uptake mutants

Changes in the two-dimensional IEF-SDS-PAGE electrophoresis (2-DE) protein profile in wild type and *trk1 trk2* mutants under high potassium (50 mM), limiting (5 mM), and starvation conditions have been analysed at the exponential and stationary phases of the cell growth curve. For that, a 2-DE coupled to MS/MS (matrix-assisted laser desorption ionization-time of flight-time of flight) strategy has been used (Curto et al., 2010; Gelis et al., 2012). With the protocol used for protein extraction around 10% of the total protein content reported for yeast as determined by the Kjeldahl method (Lagunas & Gancedo, 1973), corresponding to 40–60 μg equiv. bovine serum albumin mg^{-1} cell dry weight, could be solubilized. Between 150 and 350 spots were resolved by 2-DE in the 10–90 kDa Mr and 5–8 pI ranges. This is just a minor fraction of the whole yeast proteome indicating that most protein species are not visualized because they are not extracted and solubilized or they are present in the cell below the detection limit or out of the analysed Mr and pI ranges. In any case, our data are in agreement with those reported in the literature and show the limitations of the standard and simplest 2-DE-based proteomics. More in-depth proteome coverage requires additional steps and sample manipulation, which may lead to differences in spot resulting from manipulation artefacts and lacking biological meaning. The great advantage of the use of *Saccharomyces* as an experimental system, from the proteomics perspective, is the availability of well-curated and annotated DNA and protein databases, making the mass spectrometry (MS)-based identification quite confident with less than 20% of the protein species analysed not matching any *Saccharomyces* gene product (poor MS spectra or errors in the gene sequence). Under our experimental conditions (Curto et al., 2010), 71 gene products were identified, with most of the protein species being cytosolic and, to a lesser extent, mitochondrial or nuclear. They corresponded to enzymes of carbohydrate metabolism, being the glycolytic pathway the most represented (7 out of the 10 are involved in the conversion of glucose to pyruvate, plus alcohol dehydrogenase). Other functional groups of identified proteins corresponded to amino acid, glycerol, pyrimidine and cofactor metabolism, stress-related (antioxidants and heat shock proteins), redox homeostasis, RNA transcription, protein fate, cytoskeleton and signalling. Statistical analysis of data on spot number and intensities revealed, as occurred with cell growth estimations and potassium content parameters, more differences between growth phases than strains. Thus, principal component analysis revealed that PC1 (the highest correlation corresponded to glycolytic and redox homeostasis enzymes) and PC2 components allowed for the separation of growth phases, while for strain separation PC3 and 4 were necessary. Differences between strains were found in the stationary phase, but not during exponential growth. As a general tendency, the appearance or increase of a number of spots was observed at the stationary phase in wild type and, to a lesser extent, in mutant strains. They corresponded to enzymes involved in glycolysis and redox homeostasis, suggesting a more active glycolytic pathway in order to satisfy a higher demand of ATP and intermediates for biosynthetic routes. This higher glycolytic flux would require the activation of enzymes of methylglyoxal detoxification and NAD(P)H/NAD(P) + balance; thus, glyoxalase II and NADPH dehydrogenase were increased at the stationary phase in the wild type but not in the mutant strain, indicating an imbalance in the mutant. A more active carbohydrate catabolism would result in the overproduction of ROS, as supported by the relative accumulation of peroxiredoxin. The appearance of ubiquitin-conjugating enzymes and an increase in a number of spots in the low Mr, matching high Mr proteins, at the stationary phase in both strains indicated more active proteolysis at this stage.

Knowledge of biological processes came from comparative experiments in which different genotypes and mutants are utilized. For nutritional deficient mutants, as is the case of *S. cerevisiae trk1 trk2* cells, it is quite common to use as experimental systems wild type and derived mutant strains grown/incubated in the absence of the mutant-related nutrient (starvation conditions), in our case, K^+-free medium. Upon cell transfer to low potassium, an extraordinary decay in protein concentration and the number of 2-DE-resolved spots was observed within the first 5 h in the mutant, but not in the wild-type strain. This preceded the reported decrease in cell viability and induced programmed cell death observed 24 h after K^+ deprivation (Lauff & Santa-Maria, 2010). From these results, we can conclude that potassium starvation is a very stressful situation with probably low physiological and biological relevance when mutants were used. The increase of two ubiquitin enzymes related with DNA-repair system (Ubc2 and Mms2) support this strong stress hypothesis (Gelis et al., 2012). Under K^+-free conditions, the 2-DE protein profile of mutant and wild-type strains could be clearly differentiated, with PC1 and PC2 values of 50.7% and 22.0%, respectively, of the variability among strains and incubation times. Self-organizing maps (SOM) neural network analysis revealed a quicker adaptation to the starvation conditions for the wild-type strain (180 and 300 min samples grouped together), as compared to the double mutant (0, 30 and 60 min). During the starvation process, most of the glycolytic enzymes identified decreased in the mutant as compared with the wild-type strain. On the contrary, and according to our results (Gelis et al., 2012), other metabolic routes, such as the metabolism of amino acids and nucleotides, seemed to be unaffected.

Work is in progress in two complementary directions. First, the 2-DE protein profiles of wild type and double mutant strains grown under limiting potassium (5 mM) conditions and harvested both during exponential and stationary growth phases have been obtained. Around 300 spots were resolved in each condition. It is relevant that we found significant differences both between strains and between growth phases. MS of variable spots is under analysis. Preliminary results indicate the existence of important changes in proteins related to glycolysis. Second, our previous hypothesis will be experimentally contrasted, trying to validate proteomic data using a metabolomic approach and enzymatic analysis, focusing on carbohydrate metabolism, mainly catabolism.

5. FUNCTIONAL CONSEQUENCES OF ALTERED POTASSIUM HOMEOSTASIS

5.1. Impact on flocculation and invasiveness

The genome-wide transcriptional analysis described above and diverse research lines developed in parallel during Translucent 1 and 2 have uncovered or/and clarified unsuspected roles for potassium in yeast cells. One example is the recently reported link between potassium homeostasis and flocculation and invasiveness (Gonzalez, Casado, Petrezselyova, Ruiz, & Arino, 2013). In this case, the clue came from the previous observation that *a tetO:HAL3 vhs3* strain exhibited, under semi-restrictive conditions, a significant flocculent phenotype that was dependent on the presence of the Flo11 flocculin (Ruiz et al., 2004). Hal3 and Vhs3 are moonlighting proteins acting both as inhibitors of the Ppz protein phosphatases (de Nadal et al., 1998; Ruiz et al., 2004) and as components of a catalytic step in CoA biosynthesis (Abrie, Gonzalez, Strauss, & Arino, 2012; Ruiz et al., 2009). We demonstrated that flocculence and invasiveness in the *tetO:HAL3 vhs3* strain are caused by hyperactivation of Ppz1 as a result of depletion of its natural inhibitors. Since Ppz1 negatively controls potassium uptake through the Trk1/2 transporters (Yenush et al., 2002), hyperactivation of the phosphatase impairs potassium transport, and this results in a decrease in the intracellular pH and a subsequent increase in the levels of cAMP. Our data indicate that the signal is transmitted through the Tpk2 isoform of the catalytic subunit of protein kinase A (PKA), leading to activation of the Flo8 transcription factor, which binds to the complex *FLO11* promoter, thus modulating flocculation and invasive growth. Strains lacking the Trk1/2 transport systems show an invasive phenotype that is abolished by further deletion of *FLO8* or by increasing the potassium concentration in the medium, whereas overexpression of Trk1 attenuates *FLO11* induction in the *tetO:HAL3 vhs3* strain. It must be noted, however, that the signalling mechanism might be more complex than expected, since deletion of

RIM101 substantially reduces both expression of *FLO11* and the invasive phenotype observed in the *tetO:HAL3 vhs3* mutant. The involvement of *RIM101* in regulating *FLO11* expression and invasiveness has been reported previously (Barrales, Jimenez, & Ibeas, 2008; Li & Mitchell, 1997). However, our finding does not necessarily imply a role for Rim101 as a positive effector mediating Ppz1 signalling, since it is accepted that the Rim101 pathway is activated by alkalinization, and not by a decrease in intracellular pH. Flocculation and invasiveness are phenotypes often linked to nutrient limitation. As shown above, the presence of potassium at relatively high intracellular concentrations (200–300 mM) is required for many physiological functions. It is then conceivable that changes in the availability of this cation could be integrated by the cell in a similar way as for other nutrients.

5.2. Impact on phosphate metabolism

As mentioned above, we observed that short-term potassium deprivation results in increased mRNA levels of many *PHO*-regulated genes, such as the high-affinity phosphate transporter *PHO84*, thus mimicking a situation of phosphate starvation. A hint on a possible link between phosphate assimilation and potassium was reported as early as 1957, providing evidence that the presence of potassium in the medium was required to promote phosphate uptake (Goodman & Rothstein, 1957). However, whereas recent work has examined the impact of changes in phosphate metabolism on metal ion homeostasis (Rosenfeld et al., 2010), the inverse effect is still largely unexplored. We have confirmed the transcriptomic data by means of diverse LacZ-reporters and by evaluation of Pho84 levels by immunoblot and florescence microscopy. Induction of *PHO84* by potassium shortage is fully blocked in the absence of classic *PHO* elements like Pho81, an intracellular sensor of phosphate levels or the transcription factors Pho2 or Pho4 (D. Canadell & J. Ariño, unpublished work). Current work in our laboratory aims to understand the molecular basis linking potassium and phosphate homeostatic mechanisms.

6. MONOVALENT CATION HOMEOSTASIS: REGULATORY ISSUES

6.1. Identification of novel components regulating salt tolerance

One of the main objectives of Translucent 1 and 2 was to get further insight into the regulatory networks controlling monovalent cation homeostasis.

A relevant component of this process is the essential catalytic subunit of yeast protein phosphatase 1, termed Glc7 (Clotet, Posas, Casamayor, Schaaff-Gerstenschlager, & Arino, 1991; Feng et al., 1991). Glc7 is required for many of cellular functions and functional specificity is achieved by interaction with a variety of regulatory proteins that influence Glc7 intracellular localization and/or substrate specificity. Specific mutations in Glc7, such as that found in the *glc7-109* allele, result in salt and alkaline pH sensitivity, suggesting a role for this protein in cation homoeostasis (Williams-Hart, Wu, & Tatchell, 2002). We hypothesized that this defect could be due to a deficient interaction of the mutated protein with a regulatory subunit relevant for ion tolerance. Therefore, we screened a collection of 18 Glc7 regulatory subunit mutants for altered tolerance to diverse cations (sodium, lithium and calcium), diverse organic toxic cations and alkaline pH (Ferrer-Dalmau et al., 2010). Only deletion of *REF2* (RNA end formation 2) yielded cells sensitive to these conditions. Versions of Ref2 unable to bind Glc7 ($Ref2^{F374A}$) did not rescue the salt-related phenotypes of the *ref2* strain, indicating that Ref2 function in cation homoeostasis is mediated by Glc7. Lack of Ref2 resulted in a substantial decrease in lithium efflux, likely caused by the inability of these cells to fully induce the Na^+-ATPase *ENA1* gene. Lack of Ref2 and blockage of the calcineurin pathway yielded additive effects. *ref2* cells also display a conspicuous defect in vacuolar morphogenesis, which probably accounts for the increased calcium levels observed under standard growth conditions and the strong calcium sensitivity of this mutant. Although Ref2 has been found to be a component of the APT (associated with Pta1) subcomplex of the large CPF (cleavage and polyadenylation factor)-containing complex (holo-CPF), the collected evidence indicate that its role in salt tolerance would be unrelated to APT functions.

YPI1 encodes one of the essential regulatory subunits of Glc7, and it has been characterized as a component required for normal mitosis (Bharucha, Larson, Gao, Daves, & Tatchell, 2008; Garcia-Gimeno, Munoz, Arino, & Sanz, 2003; Pedelini et al., 2007). During the course of a parallel project, we noticed that partial depletion of Ypi1 causes lithium sensitivity, whereas high levels of this protein confer a lithium-tolerant phenotype to yeast cells (Marquina et al., 2012). Surprisingly, this phenotype was independent of the role of Ypi1 as a Glc7 regulatory subunit. We could determine that Li^+ tolerance in cells overexpressing Ypi1 was caused by a combination of effects: (i) increased efflux, due to augmented expression of the *ENA1* ATPase and (ii) decreased lithium influx through the Trk1/2 high-affinity potassium transporters. Deletion of *CNB1*, encoding the regulatory subunit of the calcineurin phosphatase, blocked Ypi1-induced expression of *ENA1*, normalized Li^+ fluxes, and abolished the Li^+ tolerance phenotype of Ypi1-overexpressing cells. Our results suggest that Ypi1, a previously defined protein phosphatase 1 regulatory subunit, controls lithium tolerance by regulating the function of a different type of protein phosphatase, calcineurin. Although this promiscuity may seem surprising, some crosstalk between regulatory subunits of different Ser/Thr protein phosphatases has been reported in the past (Castermans et al., 2012; Venturi, Bloecher, Williams-Hart, & Tatchell, 2000). Our data also suggest that Ypi1 does not impact on all calcineurin functions, suggesting that Ypi1 could contribute to the modulation of calcineurin substrate specificity.

Regulation of high-affinity potassium transport through Trk1/Trk2 has remained an elusive issue. Early work identified calcineurin as required for the transition of the transporter system to the high-affinity potassium state induced by Na^+ stress (Mendoza, Rubio, Rodriguez-Navarro, & Pardo, 1994). However, the possibility that calcineurin may play a role in the regulation of potassium transporters in the absence of stress, or the possible interaction between calcineurin and other pathways regulating Trk1 function had not been explored. During the course of Translucent 1, we demonstrated that calcineurin is necessary for high-affinity potassium uptake even in the absence of Na^+ stress (Casado et al., 2010). A pair of partially redundant protein kinases, Sat4/Hal4 and Hal5 are necessary for normal Trk1 transporter function, because they are required for maintenance of Trk1 at the plasma membrane, particularly when K^+ levels in the medium are low (Mulet et al., 1999; Perez-Valle et al., 2007). We showed that *HAL5* (but not *HAL4*) expression is induced in response to stress in a calcineurin-dependent manner through a functional calcineurin-dependent response element (CDRE) located at positions −195/−189 in the *HAL5* promoter. Lack of calcineurin decreases Hal5 protein levels, although this has little effect on the amount of Trk1 protein. However, the observation that the growth defect of *cnb1* cells at K^+-limiting conditions can be rescued in part by overexpression of *HAL5* and that this mutation further aggravates the potassium requirements of a *hal4* strain suggests that the control exerted by calcineurin on Hal5 expression may be biologically relevant for Trk1 regulation.

6.2. Pathways controlling high-pH stress response and tolerance

Budding yeast proliferates preferentially at acidic external pH. Therefore, sudden increase of the environmental pH, even barely above neutrality, represents a situation of ionic stress, and the orchestration of a proper cellular response is pivotal for survival. Work previous to Translucent, reviewed in Arino (2010), provided evidence for a complex network of signalling pathways invoked upon high pH stress, including activation of the calcineurin/Crz1, Rim101/Nrg1 and Pkc1/Slt2 pathways. However, it was evident that other response components should also exist. During Translucent, we have characterized two of these components regarding their role in high pH response: the PKA and the Snf1 pathways (Casado, Gonzalez, Platara, Ruiz, & Arino, 2011; Casamayor et al., 2012).

In-depth phenotypic analysis revealed that mutations that activate the PKA pathway tend to cause sensitivity to alkaline pH, whereas its deactivation enhances tolerance to this stress (Casado et al., 2011). We demonstrated that alkalinization causes a transient decrease in cAMP and a rapid (2–5 min) nuclear localization of the PKA-regulated Msn2 transcription factor. Msn2, together with Msn4, mediates a general stress response by binding with STRE sequences in many responsive promoters (Estruch, 2000). Our transcriptomic analysis revealed that the short-term expression of many (47%) alkaline pH-induced genes depends, at least in part, on the presence of Msn2 and Msn4. In agreement with the important role of these components, an *msn2 msn4* mutant is sensitive to alkaline pH. Therefore, this study revealed that inhibition of the PKA pathway by alkaline pH represents a substantial part of the adaptive response to this form of stress and that this response involves Msn2/Msn4-mediated genome expression remodelling.

Evidence gathered in the last few years indicated that potent and fast transcriptional response invoked by alkaline pH includes many genes repressed by glucose (Ruiz, Serrano, & Arino, 2008; Viladevall et al., 2004). Similarly, certain mutants in the glucose-sensing and response pathways, such as those lacking Snf1, the protein kinase required for adaptation to low glucose conditions, are sensitive to alkalinization. We demonstrated that the addition of glucose to the medium improves the growth of wild-type cells at high pH, fully abolishes the *snf1* alkali-sensitive phenotype and attenuates high pH-induced Snf1 phosphorylation at Thr^{210}, which is associated to activation of the Snf1 kinase (Casamayor et al., 2012). DNA microarray analysis revealed a very strong overlap (75%) between the short-term-induced genes by high pH and those induced by glucose scarcity. We found out that Snf1 mediates, in full or in part, the activation of almost 40% of the short-term alkali-induced genes (including high-affinity hexose transporters and phosphorylating enzymes, and glycogen metabolism). Genes related to trehalose metabolism, however, were largely independent of Snf1 activation, but dependent on PKA inhibition (via Msn2/4).

The results described above delineate the major signalling pathways involved in the response to high pH stress, mostly in terms of transcriptional changes. However, these studies might provide only a partial view of the problem, since they mainly focus on the impact on mRNA levels caused by promoter activation/repression. Recent work (Garcia-Martinez et al., 2012) has emphasized the relevance of mRNA stability on shaping stress responses in yeast. Therefore, we have generated (in collaboration with J. E. Pérez-Ortín, Univ. Valencia), time-resolved, genome-wide data for changes in both transcriptional rate and mRNA stability in response to alkalinization of the medium by using genomic run-on (GRO) technology. The results indicate that many short-term changes observed upon alkalinization are caused by modifications of mRNA stability, instead of promoter regulation, thus adding an additional level of complexity to this physiological response.

6.3. Modelling the regulation of the *ENA1* Na^+-ATPase gene: Experimental approaches

Modulation of gene expression has been shown to play a central role in adaptation to changing external conditions, including changes in cation composition of the medium. Diverse signalling pathways become activated, triggering the binding of transcription factors to specific DNA sequences in the promoter region, from where transcriptional responses are elicited. To be able to predict the behaviour of transcription modules, it is essential to know the biological components of a complex system and to measure *in vivo* rate and quantity parameters for each component. Surprisingly, only a few validated studies on transcriptional regulatory circuits for eukaryotic systems have been conducted so far (e.g. Janssens et al., 2006 Kim & O'Shea, 2008; Kuttykrishnan, Sabina, Langton, Johnston, & Brent, 2010). The bottleneck probably lies in experimental protocols that require careful tuning. Because rapid cell response to environmental stress is fundamental for cell survival, the Translucent 2 consortium aimed to establish a time-resolved transcriptional network model of *ENA1* Na^+/K^+ ATPase gene as an example of a highly regulated system in budding yeast that is rapidly expressed under specific stress conditions. The model comprising data defining the timing of response events will reflect the intensity of the stress.

The *ENA1* gene has received increasing attention from the time of the discovery of its cellular function (Haro, Garciadeblas, & Rodriguez-Navarro,

1991), due to its dominant role in detoxification of toxic cations by pumping sodium, lithium and surplus potassium out of the cell and in regulation of pH homeostasis. Existence of the comprehensive picture of multiple signalling pathways implicated in the regulation of the *ENA1* promoter (reviewed in Ruiz & Arino, 2007) permits the construction of a second-generation model of the *ENA1* regulatory network. Because calcineurin/Crz1 activation of *ENA1* upon alkaline stress is a well-known pathway, we sought to move beyond a steady-state description of this part of regulatory network by developing a kinetic model. Alkaline stress provokes a rapid cytosolic calcium burst that leads to execution of several steps as follows: (1) activation of Ca^{2+}/calmodulin-dependent phosphatase calcineurin, (2) dephosphorylation of the Crz1 zinc finger transcription factor, (3) subsequent translocation of Crz1 from the cytosol to the nucleus, and finally, (4) recognition and binding of Crz1 to specific DNA binding sequences, CDREs that results in induction of more than 100 different gene targets (Yoshimoto et al., 2002). The *ENA1* promoter contains two separate CDREs, of which the one localized downstream is considered to be relevant for the execution of stress-activated gene expression (Mendizabal, Pascual-Ahuir, Serrano, & de Larrinoa, 2001). The opposite activity of Crz1 is modulated by phosphorylation executed by PKA, resulting in its release from the nucleus back to the cytosol (Kafadar & Cyert, 2004). It is worth noting here the crosstalk between the calcineurin and PKA pathways (see also Section 6.2).

As mentioned above, at the first stage of second-generation modelling, we took advantage of the existing literature-based data. At the second stage, we directly measured most of the kinetic parameters for components of the regulatory network. To understand the dynamics of Crz1 localization changes, we acquired time-lapse imaging of cells expressing Crz1 fused to GFP at its native locus using a microfluidic chamber combined with fluorescence microscopy. Thus, long-term dynamic events of single cells, such as protein translocations, can be studied maintaining chemostatic conditions with time-specific flow switching capabilities. Fluorescent images were analysed using the software Cell-ID (Gordon et al., 2007) primarily tailored for yeast cells, which identifies cell boundaries and extracts fluorescent data. This approach permits tracking Crz1 within individual cells immediately upon initiation of alkaline stress conditions and observing Crz1 nuclear-localization bursts similar to those described by Cai and coworkers in response to extracellular calcium (Cai, Dalal, & Elowitz, 2008). The number of Crz1 molecules per cell has been estimated by immunoblot assay by comparison of Crz1-GFP amounts from protein extracts with known number of recombinant GFP molecules that served as internal calibration standard. To verify different affinities of both *ENA1* CDREs for Crz1 during the alkaline pH response, we employed a strain in which Crz1 was epitope-tagged with three copies of hemagglutinin (HA) at its chromosomal locus and later used in chromatin immunoprecipitation (ChiP) assay. Cell fractions were collected before and after exposure to pH 8.0 at different times, Crz1-HA proteins were cross-linked to DNA by formaldehyde, the cells were lysed and the Crz1-bound DNA was recovered by immunoprecipitation using anti-HA antibodies. The cross-linked Crz1–DNA complex was reversed and the isolated DNA was used to amplify *ENA1* promoter regions. Time-resolved quantitative binding was determined by qPCR. mRNA levels, transcription rate and mRNA half-lives data for *ENA1* were extracted from the GRO experiment of cells exposed to high pH (not published) that evaluates the transcription rate *in vivo* for all yeast genes (Garcia-Martinez, Aranda, & Perez-Ortin, 2004). Ena1 protein accumulation during alkaline stress was monitored by immunoblot of Ena1-GFP, essentially as described above for Crz1.

This work, which is near its completion, will describe the quantitative model of the *ENA1* gene regulatory network under alkaline pH stress in a time-resolved fashion. A mathematical model based on the regulatory network data will be generated and compared with model predictions. In parallel, we carried out ChiP assays for Crz1 followed by next-generation sequencing (ChiP-seq) that will allow us to study a genome-wide promoter occupancy mapping for Crz1 in response to high pH stress. It will be interesting to investigate whether Crz1 recruitment across the genome occurs in the same kinetic fashion or displays different types of temporal binding patterns. We anticipate that generation of a time-resolved transcriptional network for the *ENA1* gene under alkaline stress will establish a quantitative framework that could be broadly applied to other examples of ion and pH homeostasis regulation at the transcriptional level in other eukaryotes.

6.4. Imaging relevant components of cation homeostatic mechanisms

In a large-scale study on protein localization, most of the *S. cerevisiae* proteins were C-terminally tagged with GFP (Huh et al., 2003). This set of strains allows for the analysis of the behaviour of ion transporters and their regulators using imaging techniques. As shown in Fig. 1.8A, Nha1 and Ena1 were found mostly in the plasma membrane after growth in media with standard cation concentrations. In addition, some signal was found intracellularly.

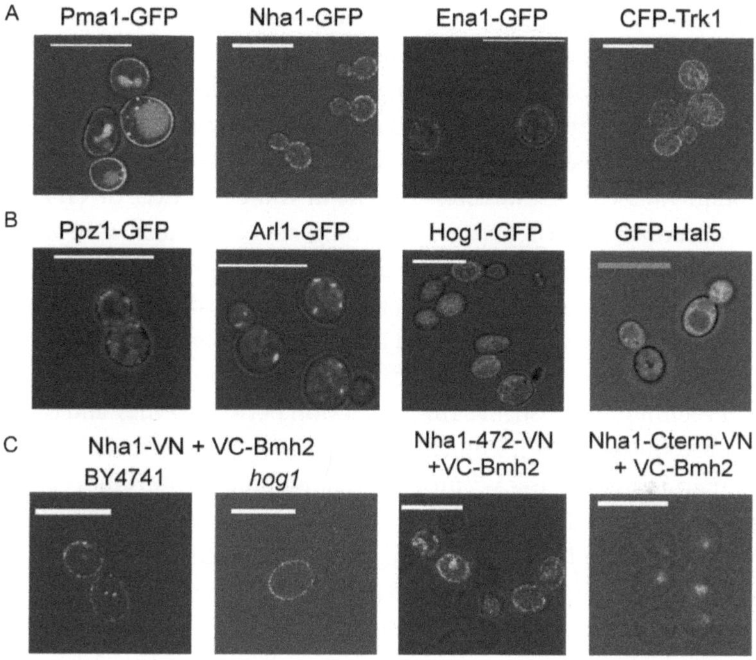

Figure 1.8 Imaging ion transporters, their regulators and interactions. (A) Confocal laser scanning images of BY4741-expressing Pma1-GFP, Nha1-GFP, Ena1-GFP or CFP-Trk1. (B) Confocal laser scanning images of BY4741-expressing Ppz1-GFP, Arl1-GFP, Hog1-GFP or GFP-Hal5. (C) Confocal laser scanning images of BY4741-expressing Nha1-VN and VC-Bmh2, of *hog1*-expressing Nha1-VN and VC-Bmh2, of BY4741-expressing $Nha1_{1\text{-}472}$-VN and VC-Bmh2 and of BY4741-expressing $Nha1_{C\text{-}term}$-VN and VC-Bmh2 (for details, see Zahradka et al., 2012). Scale bar: 10 μm. (See the color plate.)

The proton Pma1 ATPase was present both in the plasma membrane as well as in the vacuole. As a GFP-fusion of Trk1 was not present in the collection of GFP-tagged strains, we constructed a plasmid encoding an N-terminal CFP fusion of Trk1.This Trk1 was mainly found in the plasma membrane, in agreement with previous studies using C-terminally tagged Trk1 (Yenush et al., 2005). Addition of NaCl to a final concentration of 0.5 M or cultivation for 1 h in the absence of potassium did not result in an alteration of the localization of the Nha1, Ena1, Trk1 and Pma1 transporters (data not shown). Incubation for more than 4 h in the presence of 0.5 M NaCl resulted in an increase in the levels of Ena1-GFP (data not shown). We also studied the localization of protein involved in the regulation of the ion transporters. Examples of such proteins are the protein phosphatase Ppz1, the Arl1 protein and the MAP kinase Hog1. As shown in Fig. 1.8B,

Ppz1-GFP is highly enriched in the plasma membrane, making binding and regulation of Trk1 possible, and in agreement with previous data (Yenush et al., 2005). Hog1-GFP was found both in the cytoplasm and in the nucleus, whereas Arl1-GFP resides in dotted structures near the cell boundary, in agreement with a role of Arl1 in membrane trafficking. As a GFP-fusion of the protein kinase Hal5 was not present in the collection of GFP-tagged strains, we constructed a plasmid encoding an N-terminal GFP-fusion of Hal5. As shown in Fig. 1.8B, this fusion protein was present all over the cell.

Interactions between proteins can be visualized by techniques like Förster resonance energy transfer and bimolecular fluorescence complementation (BiFC) (Hu, Chinenov, & Kerppola, 2002). We successfully applied the latter technique to visualize the interaction between Nha1 and the Bmh2 14-3-3 protein (Fig. 1.8C; Zahradka et al., 2012). This interaction occurs at the cell membrane. As 14-3-3 proteins mainly bind to phosphorylated proteins and the Hog1 MAP kinase was shown to phosphorylate Nha1, we studied whether the Nha1 and Bmh2 still interact in a *hog1* mutant. However, as shown in Fig. 1.8C, *hog1* deletion did not affect the interaction. Nha1 consists of a membrane spanning N-terminal part and a large cytoplasmic C-terminal part. To investigate to which part of the Nha1 protein Bmh2 binds, we performed BiFC experiments using the N- and C-terminal part of Nha1. As shown in Fig. 1.8C, a BiFC signal was obtained in both cases, indicating that Bmh2 binds to both parts of the protein. Binding to the N-terminal part occurs in the plasma membrane as well as at intracellular membranes, whereas binding to the C-terminal part only occurs intracellularly.

6.5. Regulation of transporter biosynthesis and trafficking

One advantage of creating transporter fusions with fluorescent proteins is the ability to follow their accumulation on the subcellular level. In mammalian systems, monovalent ion transporter trafficking has long been known to play a key role in ion homeostasis regulation (reviewed in Mulet, Llopis-Torregrosa, Primo, Marques, & Yenush, 2013). For example, the ENaC Na^+ channel is known to be regulated by ubiquitylation by the HECT family E3 ubiquitin ligase, Nedd4.2. In yeast, much less is known about the impact of trafficking on ion homeostasis. As mentioned above, it is known that the Trk1 K^+ transporter is unstable at the plasma membrane in strains lacking the Hal4 and Hal5 protein kinases (Perez-Valle et al., 2007), suggesting a role for trafficking in Trk1 regulation, at least under some circumstances. The role of the GTPase Arl1 in the regulation of Trk1 further supports this hypothesis (Munson et al., 2004). In the context of Translucent, we have also observed a role for the Rim101 alkaline response pathway in the plasma membrane delivery of the Ena1 Na^+,K^+ ATPase (Marques & Yenush, unpublished data). Our results suggest that, in addition to its well-known role in transcription, a functional Rim101 complex is also required for proper delivery and/or accumulation of Ena1 at the plasma membrane in the presence of salt stress. This type of regulation is just beginning to be explored in yeast and may add yet another layer of complexity to the integrated view of the regulation of ion homeostasis under normal growth and stress conditions.

7. MATHEMATICAL MODELLING OF ION REGULATION

As discussed above, the adaptation of intracellular ion concentrations to changing extracellular conditions is based on dynamic interactions between biophysical forces and molecular regulation at the level of gene and metabolic regulation and signal transduction. Investigations of single transporters or pathways have been complemented by systems biology approaches providing a more integrated view of the coordinated responses to ionic and related stresses. We start with a brief introduction to the basic structure of the mathematical models and then summarize recent biological insights obtained from collaborative work of theoretical and wet lab biologists.

7.1. Principles of modelling ion regulation

Mathematical models of ion regulation have to reflect both the biophysical and the molecular interactions (Weiss, 1996). We use the coupling between protons and potassium (Fig. 1.9) as an example to illustrate the basic principles of modelling ion transport (Kahm et al., 2012). Extensions to more ions are then possible using the same model structure, but require more extensive experimental data and knowledge. Because of such experimental limitations, most models consider the cell as a single compartment and do not account for the dynamic distribution of ions over the various intracellular compartments. The total mass flux of potassium J_K is given as the sum over the different potassium fluxes mediated by the various potassium transport systems and causes a change of the intracellular potassium concentration

$$\frac{d[K^+]_i}{dt} = J_K. \tag{1.1}$$

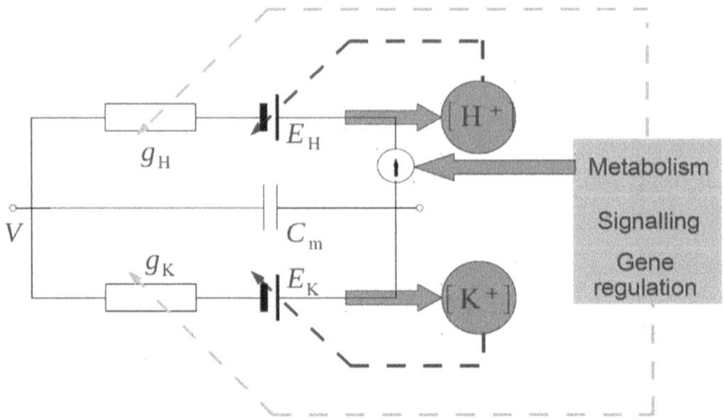

Figure 1.9 Biophysical principles of modelling ion transport using the coupling between protons and potassium as an illustrative example. Transport fluxes for potassium ions and protons change the intracellular potassium ($[K^+]_i$) and proton $[H^+]_i$ concentrations and thus the intracellular pH_i according to Eqs. (1.1) and (1.2). These concentration changes feedback to the transport fluxes via alterations of the Nernst potential E_K and E_H, via changes of the membrane voltage V and via modulations of electrical conductivities g_H and g_K of proton and potassium transport pathways. In addition, gene regulation and signal transduction control the abundance and activity of the various transport systems, what is mathematically described by increased or decreased conductivities g_H and g_K. Metabolic responses can also directly influence the ion balance by regulated injection of protons or other ions into the system.

Analogous equations have to be formulated for all other ions in the model. One essential modelling step is then to specify the flux J_K as a function of the membrane potential V and the internal and external concentrations of potassium and all other ionic species. The interaction between the different ion fluxes has been studied (Fröhlich, 2012; Gerber, 2011) with a linear nonequilibrium thermodynamics (Katchalsky & Curran, 2013) approach approximating these fluxes as linear functions of the electrochemical potentials. The nonequilibrium thermodynamics approach requires the least experimental data and does not incorporate the details of the different transport systems. To study the effect of individual transport proteins, each individual transport protein has to be modelled explicitly. In addition, leak currents for each ion mimicking the various non-specific ion currents are often included. The simple, but thermodynamically consistent relationship $J_i \propto g_i\,(V - E_i)$ relates the flux J_i of the ith transport system (protein or non-specific transport channel) to the difference of the membrane voltage V and the equilibrium potential E_i for the transported ions. The conductivity g_i can itself be a function of the membrane potential and the ion concentrations and it can also be modulated by molecular regulation mechanisms.

For example, a stress induced change of the transport activity or protein number (e.g. by phosphorylation or transcriptional regulation) of a transport system can be described by a modulation of the conductivity.

A special case has to be made for protons. In the majority of fungi and plants, the transport of alkali-metal ions, including potassium, is coupled to proton transport and thus to pH regulation. Even for the well-studied model organism *S. cerevisiae*, the details of this fundamental cell physiological process are still not sufficiently elaborated and cannot be modelled in detail (Orij, Brul, & Smits, 2011). A simple approach to effectively describe the coupling between the total flux of protons J_H across the plasma membrane to changes in intracellular pH (pH_i) is given by

$$\frac{dpH_i}{dt} = \frac{1}{\beta}(J_H + J_s), \tag{1.2}$$

where β describes the intrinsic buffering power (or buffering capacity) of the intracellular medium (Boron, 2004; Grabe & Oster, 2001). The buffering power β is actually itself a function of pH_i, but it can be approximated by a constant, since the cells interior contains a mixture of various weak acids. The number of protons required to change the pH by a fixed amount is constant over a wide pH range, due to many weak acids, the pK_a of which is distributed over a wide range. In addition to this intrinsic buffering, there are extrinsic buffering mechanisms like the CO_2/HCO^- system or other buffer pairs (Boron & De Weer, 1976). These extrinsic buffers, which are closely coupled to metabolic activities, can have an important effect on the regulation of ions and have to be accounted for by the source term J_s (Kahm et al., 2012).

Transport of ions across the plasma membrane affects not only the intracellular ion concentrations but also the corresponding Nernst potentials E_H and E_K and the membrane potential V (Fig. 1.10). This in turn has an effect on the transport activity of electrogenic transport systems and thus feeds back on J_H and J_K. Therefore, the membrane potential V has to be included as a dynamic variable. The plasma membrane is almost impermeable to polar molecules and to charged species and can therefore be seen as a capacitance C storing electrical charges $Q = CV$. These charges Q are determined by the net charges stored on a small layer on the interior and the exterior of the membrane. In practice, it is impossible to specify all these charges explicitly. For our simple potassium-proton model one can write

$$V = \frac{1}{C}\left(\beta pH_i - [K^+]_i + X\right) \tag{1.3}$$

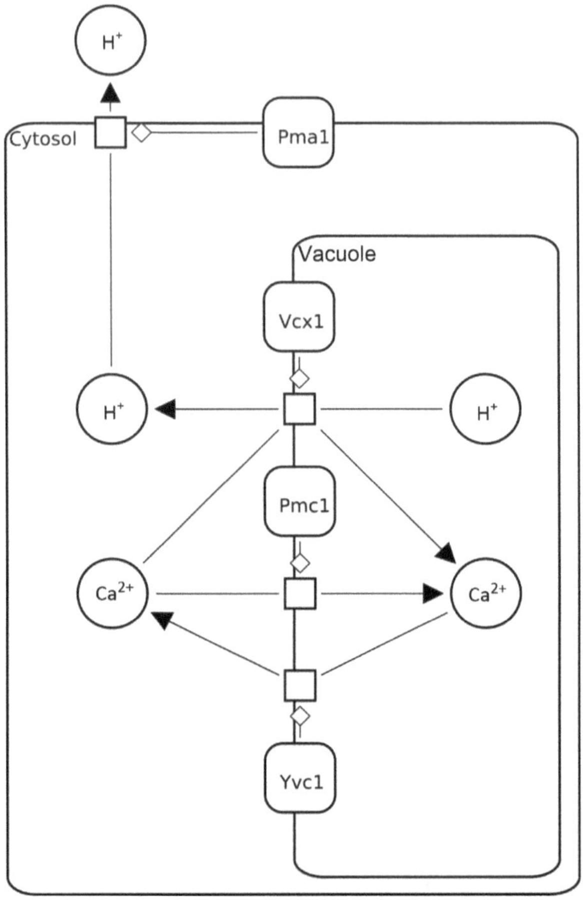

Figure 1.10 SBGN-PD visualization of the calcineurin pathway activation model created with the Biographer software. It includes the proton pump Pma1, the vacuolar antiporter Vcx1, the voltage-gated calcium channel Yvc1 and the vacuolar calcium pump Pmc1 and describes the short-peaked cytosolic calcium signal triggered by alkaline stress.

where X represents the contribution of all additional net charges to the membrane potential and constitutes an additional parameter of the model (Keener & Sneyd, 2004).

7.2. Adaptation to potassium starvation can be understood from integrated modelling and experimentation

The basic model structure sketched above was recently used to study the adaptation of *S. cerevisiae* cells to an abrupt decrease of external potassium concentrations (Kahm et al., 2012). In this work, the total potassium flux J_K is given by the sum over transport fluxes across the Trk1,2 system, the Tok1 channel and the Nha1 antiporter. In addition, an equivalent potassium flux representing other, mainly non-specific transport systems was included. The proton flux J_H incorporates the pumping of the H^+-ATPase Pma1, proton transport across the Nha1 antiporter, and an equivalent proton flux accounting for the effect of the various secondary transport mechanisms coupled to protons. The Ena1 ATPase was not included in this model, because the starvation experiments used to calibrate the model were performed at medium to low external pH, where this protein has only low transport activity (Ruiz & Arino, 2007).

A major challenge in model development was the explicit description of the concentration- and voltage-dependent kinetics of the various transport proteins. For Trk1, Trk2 and Tok1, electrophysiological data from the literature (Johansson & Blatt, 2006; Kuroda, Bihler, Bashi, Slayman, & Rivetta, 2004) were used to estimate the voltage- and concentration-dependent conductivities and a description of the Pma1 ATPase was based on simple thermodynamic considerations. All other transport systems were described by the simplest possible thermodynamic relationships as described in Section 7.1. By comparing experimentally observed steady-state levels for potassium with the model simulations, we found that the maintenance of a minimum intracellular potassium concentration is only possible when the Nha1 transporter decreases its activity upon starvation. Therefore, this transporter was assumed to decrease its conductivity with increasing (more negative) membrane voltage (Ohgaki et al., 2005). All other conductivities in the model were initially assumed to be constant parameters.

The inference of free parameters from experimental time course data is a critical step in developing mechanistic systems biology models. By applying a variety of computational methods for parameter estimation (Raue et al., 2009; Zi, 2011), we were able to reproduce the experimental steady-state concentrations of intracellular potassium, but not the dynamics of the intracellular potassium concentrations. In the model, the dynamics of potassium loss was always much faster than in the experiments. This failure to reproduce a relatively simple potassium starvation experiment was an indication that an important physiological mechanism required for the adaptation to the very low external potassium concentrations was not considered in the model. In principle, this unmodelled dynamic process could have been everything from signalling to metabolic responses. To overcome this problem, we devised a new method, the Reverse Tracking Algorithm (RTA).

It is a control theoretic method that suggests dynamic model modifications that allow the model to reproduce the data and excludes model modifications, which are not supported by the data. The suggested modifications can then be used to design new experiments, which test whether the predicted change is indeed happening in biology.

One modification predicted by the RTA was the activation of the Pma1 ATPase and an increased reaction rate for the bicarbonate reaction (related to the source flux J_s in Eq. 1.2). The activation of Pma1 was experimentally verified by measuring Pma1 activity from crude membrane preparations. Experimental support for the transient activation of the bicarbonate reaction system in response to potassium starvation came from gene expression measurements of the gene *NCE103* coding for carbonic anhydrase. The regulated activation of proton fluxes is an essential mechanism to keep intracellular potassium concentrations sufficiently high. This view, as obtained from modelling and model-driven experimentation, shows that potassium regulation requires more than a coordinated opening and closing of transport channels and provides a further hint that responses to low potassium have an effect of metabolic activities, which in turn are used for the adaptation to low potassium. The detailed mechanisms, how these increased proton fluxes are triggered and how the cell senses the loss of intracellular potassium are still to be elucidated.

7.3. Computational modelling of the Ena1 Na^+,K^+-ATPase activity

The model by Kahm et al. (2012) does not consider the *ENA1* ATPase, since it was used to analyse the responses to an abrupt decrease of the extracellular potassium content under conditions of a relatively low pH. However, responses to rapid changes of the extracellular NaCl concentration or pH (see Section 6.3; Ruiz & Arino, 2007) can only be understood, when *ENA1* regulation is explicitly modelled. Initially, the only information available was given by a wiring diagram describing the interaction of the *ENA1* promoter (Arino, 2010; Ruiz & Arino, 2007) with various signalling pathways including the calcineurin/Crz1 pathway. This qualitative information was translated to a Boolean model describing the *ENA1* promoter activity as either "off" or "on", depending on the binary states of the various activating or inhibiting signalling pathways. The new modelling framework rxncon (Tiger et al., 2012) was helpful for model development and made possible the transformation of this qualitative Boolean model to an ordinary differential equation model describing the interactions between the calcineurin pathway and the *ENA1* promoter. This model is still under development and we are currently integrating the data described in Section 6.3 for quantitative promoter binding, mRNA levels and mRNA half lives into the model. Motivated by data described by Denis and Cyert (2002), we also explored the possibility that a release of calcium from the vacuole is caused by the depolarization of the vacuolar membrane, which in turn is caused by an imbalance of ions between the cytosol and the extracellular space. This could be a potential mechanism for the stimulation of *ENA1* expression in response to an increased cytosolic calcium concentration, which is transmitted by the calcineurin pathway. Accordingly, further vacuolar transport systems might be involved in shaping of this signal under alkaline pH stress or NaCl stress. This model is visualized in Fig. 1.10 using the System Biology Graphical Notation language Process Description created with the recently published biographer software (Krause et al., 2013).

7.4. Modelling of other ion species and metabolic effects

In cases where significant changes of the cell volume are experimentally observed, the flux and membrane potential equations have to be complemented by further equations for volume changes and by descriptions of the HOG and calcineurin pathways essential for osmotic pressure response (Schaber, Baltanas, Bush, Klipp, & Colman-Lerner, 2012). Recently, the cellular responses to NaCl, sorbitol, KCl and alkaline pH stress were studied by means of a model adding sodium cations to the above described model (Ke, Ingram, & Haynes, 2013). As a result, it was predicted that the interaction between Hog1 and Tok1 at the plasma membrane inhibits Tok1 and reduces Na^+ influx under NaCl stress. Additionally, the calcineurin pathway seems to be important for maintaining sufficiently low levels of Na^+ in the long-term response to NaCl stress, but not for the control of Na^+ under the other stresses investigated. Fitting the parameters of such large dynamical models based on mostly static measurements is an almost intractable challenge. As a result, it is often the case that only a subpart of the observations are well reproduced. For instance, an inconsistency of the model is that increased activity of the Trk1,2 transporters depolarizes the plasma membrane and thus increases the Na^+/K^+ ratio, which contradicts the experimental findings. The authors attribute this to the simplified representation of the Trk1,2 system in the model. Motivated by the results reviewed in Section 7.2, it could also be speculated that a modulation of proton fluxes could counteract the depolarization of the membrane potential, which could be tested experimentally by measuring the activity of Pma1 and of the bicarbonate reaction (Kahm et al., 2012).

Recent data (Herrera et al., 2013) could provide a basis for future models studying the interaction of the biophysical and regulatory forces responsible for the distribution of ions over different intracellular compartments using the basic model structure described in Section 7.1 for each compartment. A major challenge for such a model is that the kinetic characterization of transport proteins for the various compartments is still difficult or impossible. In addition, the membrane potential between the compartment and the cytosol is not easily accessible experimentally.

Experimental findings consistently show that ion regulation and various metabolic pathways are tightly coupled. True integrated modelling will require the incorporation of a metabolism module describing metabolic adjustments in response to ionic or related stresses. For medium- and long-term responses, the short relaxation time scales for metabolic reactions justify a quasi-steady-state approximation for this metabolism module. Genome-wide metabolic network reconstructions for *S. cerevisiae* are available (Forster, Famili, Fu, Palsson, & Nielsen, 2003) and constantly improving (Heavner, Smallbone, Barker, Mendes, & Walker, 2012). Metabolic simulations including flux balance analysis and related methods can be used in conjunction with large-scale gene expression, proteome or metabolome data to build context-specific metabolic models for comparing metabolic flux states under different stress conditions (Lewis, Nagarajan, & Palsson, 2012). A first attempt in this direction is described in Mo, Palsson, and Herrgard (2009), where internal metabolic states under conditions of low potassium are compared to conditions of excess ammonia. Such models provide a genome-wide overview of steady-state metabolic adjustments and can also predict growth rates and the effect of gene or reaction knock outs under various conditions, which can readily be compared to experimental data (Lewis et al., 2012).

Large-scale modelling of transient metabolic or other responses is currently limited to smaller sub-networks or processes. Sometimes, explicit modelling might be worthwhile for central parts of ion regulation, but sometimes an approach handling regulatory processes as dynamic external inputs using advanced modelling tools for handling unmodelled dynamics might be easier and sufficient for integrating models and modules for different time scales.

8. DATA MANAGEMENT

8.1. Data storage in systems biology

A key method of systems biology is model-driven experimentation. This process requires efficient data storage, exchange and management. Exchange of spreadsheets via email attachments is clearly not sufficient for integrated and reproducible research. To establish a new culture of data sharing, user friendly software tools are essential, because otherwise scientists will fall back into their old habits of keeping their data locally. Our initial attempt to create a suitable Web-based data management solution was based on the open source group management software PhProjekt. However, it turned out that the extension and modification of this system was too labour intensive. Therefore, we decided to design our own data management platform called pyMantis based on the web framework web2py. Its main features are: relational data storage, tree-based file storage, HTML file previews (among others: .xls, .odt, .odf, .doc, .pdf) annotations, search, wiki and a web service interface (using a RESTful API; Fielding, 2000).

Our primary goal for the data management of the Translucent project was to create a relational data storage accessible via a Web-based interface. We, however, faced the major challenge of data management in biology: the type and format of the data to be exchanged cannot be defined from the beginning and are subject to frequent changes. While there are attempts that facilitate a top-down data modelling approach (Rocca-Serra et al., 2010; Wolstencroft et al., 2011), the experience we got from our first platform showed us that we needed a bottom-up data modelling approach. In order to achieve this, we used the abilities of web2py to create a user interface for creating and modifying our relational data models as well as the user interfaces and web services in the running platform while still enabling the data insertion and manipulation in parallel. We were able to do this by storing the information about our data model and user interface in the SQL database and creating an algorithm that writes the source code for the object relational mapper of web2py. This source code is then interpreted by HTML rendering algorithms as well as the migration algorithm (Active Record Migrations—Ruby on Rails Guides, http://guides.rubyonrails.org/migrations.html) of web2py that can adapt the SQL data tables based on the current data model. With this, we were able to reduce the time span of creating or modifying the relational data storage, including all interfaces, to a few minutes. The fast adaption of the platform was one of the points that lead to a high user acceptance.

8.2. Best practices in data storage

In the SysMO project, the SysMO-DB consortium was established to coordinate the data management of all consortia. The SysMO-DB consortium introduced a group of data management representatives (PALs), including theoreticians and biologists from each consortium, which held biannual meetings. Due to the communication within the PALs group, we were able to establish and enforce several best practices in data storage:

- All data must contain a set of basic metadata: identifier, human readable identifier, creator, editor, created on, edited on, version.
- All data must adhere to minimum information criteria (Taylor et al., 2008). That mean that a dataset can only be stored if a list of predefined information is provided. (We additionally check for correct data types.)
    - All data should be categorized in a three layered textual description scheme introduced by the ISA-tool group (Baker et al., 2013)
    - Investigation: the project context
    - Study: a unit of research
    - Assay: analytical measurement
- All data files must be available under a stable URL identifier. That means that a new version of the same data file will be accessible under the same URL with a hint to the current version and a method for accessing the older version.
- No data will ever be deleted, only hidden.
- Data should be accessible via web services for further access.

8.3. Data sharing policies

A key point in the discussion about the data sharing was the access management for the data. We implemented our access management based on role-based access control. That means that we created groups that each user belongs to, each group is then ascribed certain default access rights. While it was also possible for users to create their own groups and access policies the default access rights were in almost all cases accepted and used by all consortia members. Our default access policies were:

- all uploaded data was readable to all Translucent members
- only the meta data (see Section 8.2) are shared with the whole SysMO project

- external visitors have no right to view the data
- the data and its access rights can only be changed by the creator of a dataset (and the editor if different from the creator)

This meant that sharing the data beyond our own consortia was encouraged, but required the interaction of the creator of a dataset.

As data sharing is one of the goals of the SysMO project, it was furthermore agreed on by all consortia that within a certain time span after the end of the SysMO project all collected data will be made publicly available.

At the end of our project, we can state that our data management was well accepted and used. While some methods of communication were not accepted by its users (forum, calendar and chat) other methods were frequently used. This includes the tree base file storage, which was used to collect presentations of our consortium meetings among other documents, with around 350 documents collected. Also the relational data storage were well accepted and we were able to collect almost 1000 datasets. The reasons for its success are many-fold and include technical aspects, like being able to adapt quickly to new needs, or having a templating system that facilitates the insertion of large amounts of data, but also includes non-technical aspects like the adaptation of a visual design strategy by applying typography techniques like grid layouts, as well as social organization strategies which meant that for example the collection of the presentations was carried out by one dedicated consortium member directly after a meeting.

To create an even better platform, we are attempting to generate a Web-based mailing list interface, helping to centralize all email communication. We are furthermore working on algorithms to create spreadsheet-based templates that can be automatically parsed and inserted into our relational storage, as well as algorithms for automatically extracting data from arbitrary spreadsheets.

The data management platform pyMantis is available as an open source project (the source code can be found at http://sourceforge.net/projects/tlc2/) with packages for different operating systems available (Linux, OsX and Windows; at http://pymantis.org).

9. CONCLUSIONS

The coordinated work of the different partners involved in Translucent 1 and 2 has substantially improved our knowledge on monovalent cation homeostasis in yeast and its influence in overall yeast biology. We have generated novel tools, reagents and methodologies that will be useful for future projects, such as dozens of strains carrying combinations of mutations of interest, or tagged versions of relevant proteins, or plasmids to visualize protein–protein interactions by the BiFC technique. Importantly, the definition of standard experimental conditions and SOPs was a key point to allow for the integration of the data. We also developed new methods for measurement of subcellular cation content, for time-resolved determination of ion fluxes across the yeast's plasma membrane (FLISE), and for quantitative analyses of membrane potential changes.

Work during Translucent has produced a great deal of data, as reflected in the nearly 600 assets deposited in the database. These include genome-wide screens for altered tolerance to toxic cationic drugs and for potassium requirements, transcriptomic, metabolomic and proteomic studies on the effects of potassium starvation, time-resolved determination of cation fluxes in wild type and relevant mutants, study of the interaction of 14-3-3 proteins with key components of the homeostatic mechanisms, etc. This information has been completed with the discovery of new regulatory mechanisms of monovalent cation influx and efflux and novel functional interactions between these mechanisms and other cellular functions, such as flocculation and invasiveness. Therefore, we have now a much wider (and also more detailed) view of the cellular relevance of monovalent cation homeostasis than we had at the beginning of the project.

Data collection was integrated with the development of the first mathematical model of potassium control in *S. cerevisiae*, revealing that the regulation of proton fluxes by activation of the Pma1 ATPase and the bicarbonate reaction are key events for potassium homeostasis. In addition, the mechanisms involved in the regulation of stable levels of intracellular potassium combining the function of different active and passive cation transporters were elucidated. An important methodological progress underlying this model was the development of the reverse tracking algorithm (RTA), which allows for a systematic data-driven identification of unmodelled dynamic processes in systems biological models.

Pivotal to the development of the project was the design and construction of the pyMantis data management platform that integrates several of the most common methods of data sharing and digital communication: wiki, email, tree-based file/folder structure, relational database, annotations/tags, ISA data categorization, web services for common data exchange formats, and many more.

The continuous interaction between modellers and experimentalists (two general meetings every year, plus numerous bi- and tri-lateral meetings) allowed us to successfully establish a feedback mechanism that is at the basis of systems biology studies. From the modeller's side, it was pivotal to have a very detailed understanding of the biology and to know what the interesting biological questions were and how difficult it could be to generate the required data. This helped the experimentalist groups to set up the necessary experimental conditions.

Translucent 1 and 2 have generated a substantial number of publications (33 at the moment of writing these lines) and are expected to produce several more papers in the next few months. Importantly, many of these papers are joint papers, highlighting the interactions among consortium members. Besides this generation of knowledge, Translucent has been important because the interaction between the different groups, with diverse backgrounds and expertise, has enriched all of us with a new and wider perspective of yeast biology.

ACKNOWLEDGEMENTS

The partners of Translucent wish to acknowledge the effort of the SysMO Steering Committee, SysMO Office and Advisory Board during all this period. J. A. wishes to thank Julio Barbas and José A. Salas for their enthusiasm and help. Work supported by grants GEN2006-27748-C2-1 and EUI2009-04147 (Ministerio de Ciencia e Innovación, Spain, and FEDER) to J. A.; grants 826.06.004 and 826.09.006 (Netherlands Organization for Scientific Research (NWO)—Earth and Life Sciences (ALW)) to G. P. H. v. H.; grants GA CR P503/10/0307, GA AS CR IAA500110801 and MSMT COST LD13037 to H. S. and BBSRC grants to C. P.

REFERENCES

Abrie, J. A., Gonzalez, A., Strauss, E., & Arino, J. (2012). Functional mapping of the disparate activities of the yeast moonlighting protein Hal3. *The Biochemical Journal*, *442*, 357–368.

Aguilera, J., & Prieto, J. A. (2004). Yeast cells display a regulatory mechanism in response to methylglyoxal. *FEMS Yeast Research*, *4*, 633–641.

Arino, J. (2010). Integrative responses to high pH stress in S. cerevisiae. *OMICS*, *14*, 517–523.

Arino, J., Ramos, J., & Sychrova, H. (2010). Alkali metal cation transport and homeostasis in yeasts. *Microbiology and Molecular Biology Reviews*, *74*, 95–120.

Baker, N. A., Klemm, J. D., Harper, S. L., Gaheen, S., Heiskanen, M., Rocca-Serra, P., et al. (2013). Standardizing data. *Nature Nanotechnology*, *8*, 73–74.

Barrales, R. R., Jimenez, J., & Ibeas, J. I. (2008). Identification of novel activation mechanisms for FLO11 regulation in Saccharomyces cerevisiae. *Genetics*, *178*, 145–156.

Barreto, L., Canadell, D., Petrezselyova, S., Navarrete, C., Maresova, L., Perez-Valle, J., et al. (2011). A genomewide screen for tolerance to cationic drugs reveals genes important for potassium homeostasis in Saccharomyces cerevisiae. *Eukaryotic Cell*, *10*, 1241–1250.

Barreto, L., Canadell, D., Valverde-Saubi, D., Casamayor, A., & Arino, J. (2012). The short-term response of yeast to potassium starvation. *Environmental Microbiology*, *14*, 3026–3042.

Bertl, A., Bihler, H., Kettner, C., & Slayman, C. L. (1998). Electrophysiology in the eukaryotic model cell Saccharomyces cerevisiae. *Pflügers Archiv, 436*, 999–1013.

Bertl, A., Bihler, H., Reid, J. D., Kettner, C., & Slayman, C. L. (1998). Physiological characterization of the yeast plasma membrane outward rectifying K^+ channel, DUK1 (TOK1), in situ. *The Journal of Membrane Biology, 162*, 67–80.

Bertl, A., Ramos, J., Ludwig, J., Lichtenberg-Frate, H., Reid, J., Bihler, H., et al. (2003). Characterization of potassium transport in wild-type and isogenic yeast strains carrying all combinations of trk1, trk2 and tok1 null mutations. *Molecular Microbiology, 47*, 767–780.

Bertl, A., & Slayman, C. L. (1990). Cation-selective channels in the vacuolar membrane of Saccharomyces: Dependence on calcium, redox state, and voltage. *Proceedings of the National Academy of Sciences of the United States of America, 87*, 7824–7828.

Bharucha, J. P., Larson, J. R., Gao, L., Daves, L. K., & Tatchell, K. (2008). Ypi1, a positive regulator of nuclear protein phosphatase type 1 activity in Saccharomyces cerevisiae. *Molecular Biology of the Cell, 19*, 1032–1045.

Bihler, H., Slayman, C. L., & Bertl, A. (1998). NSC1: A novel high-current inward rectifier for cations in the plasma membrane of Saccharomyces cerevisiae. *FEBS Letters, 432*, 59–64.

Boron, W. F. (2004). Regulation of intracellular pH. *Advances in Physiology Education, 28*, 160–179.

Boron, W. F., & De Weer, P. (1976). Intracellular pH transients in squid giant axons caused by CO_2, NH_3, and metabolic inhibitors. *The Journal of General Physiology, 67*, 91–112.

Brett, C. L., Tukaye, D. N., Mukherjee, S., & Rao, R. (2005). The yeast endosomal Na^+K+/H^+ exchanger Nhx1 regulates cellular pH to control vesicle trafficking. *Molecular Biology of the Cell, 16*, 1396–1405.

Cagnac, O., Leterrier, M., Yeager, M., & Blumwald, E. (2007). Identification and characterization of Vnx1p, a novel type of vacuolar monovalent cation/H^+ antiporter of Saccharomyces cerevisiae. *The Journal of Biological Chemistry, 282*, 24284–24293.

Cai, L., Dalal, C. K., & Elowitz, M. B. (2008). Frequency-modulated nuclear localization bursts coordinate gene regulation. *Nature, 455*, 485–490.

Casado, C., Gonzalez, A., Platara, M., Ruiz, A., & Arino, J. (2011). The role of the protein kinase A pathway in the response to alkaline pH stress in yeast. *The Biochemical Journal, 438*, 523–533.

Casado, C., Yenush, L., Melero, C., Ruiz, M. D., Serrano, R., Perez-Valle, J., et al. (2010). Regulation of Trk-dependent potassium transport by the calcineurin pathway involves the Hal5 kinase. *FEBS Letters, 584*, 2415–2420.

Casamayor, A., Serrano, R., Platara, M., Casado, C., Ruiz, A., & Arino, J. (2012). The role of the Snf1 kinase in the adaptive response of Saccharomyces cerevisiae to alkaline pH stress. *The Biochemical Journal, 444*, 39–49.

Castermans, D., Somers, I., Kriel, J., Louwet, W., Wera, S., Versele, M., et al. (2012). Glucose-induced posttranslational activation of protein phosphatases PP2A and PP1 in yeast. *Cell Research, 22*, 1058–1077.

Cline, M. S., Smoot, M., Cerami, E., Kuchinsky, A., Landys, N., Workman, C., et al. (2007). Integration of biological networks and gene expression data using Cytoscape. *Nature Protocols, 2*, 2366–2382.

Clotet, J., Posas, F., Casamayor, A., Schaaff-Gerstenschlager, I., & Arino, J. (1991). The gene DIS2S1 is essential in Saccharomyces cerevisiae and is involved in glycogen phosphorylase activation. *Current Genetics, 19*, 339–342.

Curto, M., Valledor, L., Navarrete, C., Gutierrez, D., Sychrova, H., Ramos, J., et al. (2010). 2-DE based proteomic analysis of Saccharomyces cerevisiae wild and K^+ transport-affected mutant (trk1,2) strains at the growth exponential and stationary phases. *Journal of Proteomics, 73*, 2316–2335.

Cyert, M. S., & Philpott, C. C. (2013). Regulation of cation balance in Saccharomyces cerevisiae. *Genetics*, *193*, 677–713.

de Nadal, E., Clotet, J., Posas, F., Serrano, R., Gomez, N., & Arino, J. (1998). The yeast halotolerance determinant Hal3p is an inhibitory subunit of the Ppz1p Ser/Thr protein phosphatase. *Proceedings of the National Academy of Sciences of the United States of America*, *95*, 7357–7362.

Denis, V., & Cyert, M. S. (2002). Internal Ca(2+) release in yeast is triggered by hypertonic shock and mediated by a TRP channel homologue. *The Journal of Cell Biology*, *156*, 29–34.

Desai, K. M., Chang, T., Wang, H., Banigesh, A., Dhar, A., Liu, J., et al. (2010). Oxidative stress and aging: Is methylglyoxal the hidden enemy? *Canadian Journal of Physiology and Pharmacology*, *88*, 273–284.

Eraso, P., Mazon, M. J., & Portillo, F. (2006). Yeast protein kinase Ptk2 localizes at the plasma membrane and phosphorylates in vitro the C-terminal peptide of the H^+ − ATPase. *Biochimica et Biophysica Acta*, *1758*, 164–170.

Eraso, P., & Portillo, F. (1994). Molecular mechanism of regulation of yeast plasma membrane H(+)-ATPase by glucose. Interaction between domains and identification of new regulatory sites. *The Journal of Biological Chemistry*, *269*, 10393–10399.

Estruch, F. (2000). Stress-controlled transcription factors, stress-induced genes and stress tolerance in budding yeast. *FEMS Microbiology Reviews*, *24*, 469–486.

Feng, Z. H., Wilson, S. E., Peng, Z. Y., Schlender, K. K., Reimann, E. M., & Trumbly, R. J. (1991). The yeast GLC7 gene required for glycogen accumulation encodes a type 1 protein phosphatase. *The Journal of Biological Chemistry*, *266*, 23796–23801.

Ferrer-Dalmau, J., Gonzalez, A., Platara, M., Navarrete, C., Martinez, J. L., Barreto, L., et al. (2010). Ref2, a regulatory subunit of the yeast protein phosphatase 1, is a novel component of cation homoeostasis. *The Biochemical Journal*, *426*, 355–364.

Fielding, R. T. (2000). *Architectural styles and the design of network-based software*. PhD Dissertation, Irvine, CA: University of California.

Forster, J., Famili, I., Fu, P., Palsson, B. O., & Nielsen, J. (2003). Genome-scale reconstruction of the Saccharomyces cerevisiae metabolic network. *Genome Research*, *13*, 244–253.

Fröhlich, M. (2012). *Computational and experimental investigation of the mechanisms involved in cellular water transport and (cat)ion homeostasis*. PhD Dissertation, Berlin: Freie Universität Berlin.

Garciadeblas, B., Rubio, F., Quintero, F. J., Banuelos, M. A., Haro, R., & Rodriguez-Navarro, A. (1993). Differential expression of two genes encoding isoforms of the ATPase involved in sodium efflux in Saccharomyces cerevisiae. *Molecular and General Genetics*, *236*, 363–368.

Garcia-Gimeno, M. A., Munoz, I., Arino, J., & Sanz, P. (2003). Molecular characterization of Ypi1, a novel *Saccharomyces cerevisiae* type 1 protein phosphatase inhibitor. *The Journal of Biological Chemistry*, *278*, 47744–47752.

Garcia-Martinez, J., Aranda, A., & Perez-Ortin, J. E. (2004). Genomic run-on evaluates transcription rates for all yeast genes and identifies gene regulatory mechanisms. *Molecular Cell*, *15*, 303–313.

Garcia-Martinez, J., Ayala, G., Pelechano, V., Chavez, S., Herrero, E., & Perez-Ortin, J. E. (2012). The relative importance of transcription rate, cryptic transcription and mRNA stability on shaping stress responses in yeast. *Transcription*, *3*, 39–44.

Gelis, S., Curto, M., Valledor, L., Gonzalez, A., Arino, J., Jorrin, J., et al. (2012). Adaptation to potassium starvation of wild-type and K(+)-transport mutant (trk1,2) of Saccharomyces cerevisiae: 2-Dimensional gel electrophoresis-based proteomic approach. *Microbiology Open*, *1*, 182–193.

Gerber, S. (2011). *In silico modeling of cation homeostasis in* Saccharomyces cerevisiae. PhD Thesis, Berlin: Humboldt Universität zu Berlin.

Gonzalez, A., Casado, C., Petrezselyova, S., Ruiz, A., & Arino, J. (2013). Molecular analysis of a conditional hal3 vhs3 yeast mutant links potassium homeostasis with flocculation and invasiveness. *Fungal Genetics and Biology*, *53*, 1–9.

Goodman, J., & Rothstein, A. (1957). The active transport of phosphate into the yeast cell. *The Journal of General Physiology*, *40*, 915–923.

Goossens, A., de La Fuente, N., Forment, J., Serrano, R., & Portillo, F. (2000). Regulation of yeast H(+)-ATPase by protein kinases belonging to a family dedicated to activation of plasma membrane transporters. *Molecular and Cellular Biology*, *20*, 7654–7661.

Gordon, A., Colman-Lerner, A., Chin, T. E., Benjamin, K. R., Yu, R. C., & Brent, R. (2007). Single-cell quantification of molecules and rates using open-source microscope-based cytometry. *Nature Methods*, *4*, 175–181.

Grabe, M., & Oster, G. (2001). Regulation of organelle acidity. *The Journal of General Physiology*, *117*, 329–344.

Guldener, U., Heck, S., Fielder, T., Beinhauer, J., & Hegemann, J. H. (1996). A new efficient gene disruption cassette for repeated use in budding yeast. *Nucleic Acids Research*, *24*, 2519–2524.

Gustin, M. C., Zhou, X. L., Martinac, B., & Kung, C. (1988). A mechanosensitive ion channel in the yeast plasma membrane. *Science*, *242*, 762–765.

Haro, R., Garciadeblas, B., & Rodriguez-Navarro, A. (1991). A novel P-type ATPase from yeast involved in sodium transport. *FEBS Letters*, *291*, 189–191.

Haro, R., & Rodriguez-Navarro, A. (2002). Molecular analysis of the mechanism of potassium uptake through the TRK1 transporter of Saccharomyces cerevisiae. *Biochimica et Biophysica Acta*, *1564*, 114–122.

Haro, R., & Rodriguez-Navarro, A. (2003). Functional analysis of the M2(D) helix of the TRK1 potassium transporter of Saccharomyces cerevisiae. *Biochimica et Biophysica Acta*, *1613*, 1–6.

Heavner, B. D., Smallbone, K., Barker, B., Mendes, P., & Walker, L. P. (2012). Yeast 5—An expanded reconstruction of the Saccharomyces cerevisiae metabolic network. *BMC Systems Biology*, *6*, 55.

Herrera, R., Alvarez, M. C., Gelis, S., & Ramos, J. (2013). Subcellular potassium and sodium distribution in Saccharomyces cerevisiae wild type and vacuolar mutants. *The Biochemical Journal*, *454*, 525–532.

Hess, D. C., Lu, W., Rabinowitz, J. D., & Botstein, D. (2006). Ammonium toxicity and potassium limitation in yeast. *PLoS Biology*, *4*, e351.

Hoeberichts, F. A., Perez-Valle, J., Montesinos, C., Mulet, J. M., Planes, M. D., Hueso, G., et al. (2010). The role of K(+) and H(+) transport systems during glucose- and H(2)O(2)-induced cell death in Saccharomyces cerevisiae. *Yeast*, *27*, 713–725.

Hu, C. D., Chinenov, Y., & Kerppola, T. K. (2002). Visualization of interactions among bZIP and Rel family proteins in living cells using bimolecular fluorescence complementation. *Molecular Cell*, *9*, 789–798.

Huh, W. K., Falvo, J. V., Gerke, L. C., Carroll, A. S., Howson, R. W., Weissman, J. S., et al. (2003). Global analysis of protein localization in budding yeast. *Nature*, *425*, 686–691.

Janssens, H., Hou, S., Jaeger, J., Kim, A. R., Myasnikova, E., Sharp, D., et al. (2006). Quantitative and predictive model of transcriptional control of the Drosophila melanogaster even skipped gene. *Nature Genetics*, *38*, 1159–1165.

Johansson, I., & Blatt, M. R. (2006). Interactive domains between pore loops of the yeast K^+ channel TOK1 associate with extracellular K^+ sensitivity. *The Biochemical Journal*, *393*, 645–655.

Kafadar, K. A., & Cyert, M. S. (2004). Integration of stress responses: Modulation of calcineurin signaling in Saccharomyces cerevisiae by protein kinase A. *Eukaryotic Cell*, *3*, 1147–1153.

Kahm, M., Navarrete, C., Llopis-Torregrosa, V., Herrera, R., Barreto, L., Yenush, L., et al. (2012). Potassium starvation in yeast: Mechanisms of homeostasis revealed by mathematical modeling. *PLoS Computational Biology*, *8*, e1002548.

Kanzaki, M., Nagasawa, M., Kojima, I., Sato, C., Naruse, K., Sokabe, M., et al. (1999). Molecular identification of a eukaryotic, stretch-activated nonselective cation channel. *Science*, *285*, 882–886.

Katchalsky, A., & Curran, P. (2013). *Nonequilibrium thermodynamics in biophysics*. Cambridge: Harvard University Press.

Ke, R., Ingram, P. J., & Haynes, K. (2013). An integrative model of ion regulation in yeast. *PLoS Computational Biology*, *9*, e1002879.

Keener, J., & Sneyd, J. (2004). *Mathematical physiology*. New York: Springer.

Ketchum, K. A., Joiner, W. J., Sellers, A. J., Kaczmarek, L. K., & Goldstein, S. A. (1995). A new family of outwardly rectifying potassium channel proteins with two pore domains in tandem. *Nature*, *376*, 690–695.

Kim, H. D., & O'Shea, E. K. (2008). A quantitative model of transcription factor-activated gene expression. *Nature Structural and Molecular Biology*, *15*, 1192–1198.

Kinclova, O., Ramos, J., Potier, S., & Sychrova, H. (2001). Functional study of the Saccharomyces cerevisiae Nha1p C-terminus. *Molecular Microbiology*, *40*, 656–668.

Krause, F., Schulz, M., Ripkens, B., Flottmann, M., Krantz, M., Klipp, E., et al. (2013). Biographer: Web-based editing and rendering of SBGN compliant biochemical networks. *Bioinformatics*, *29*, 1467–1468.

Kuroda, T., Bihler, H., Bashi, E., Slayman, C. L., & Rivetta, A. (2004). Chloride channel function in the yeast TRK-potassium transporters. *The Journal of Membrane Biology*, *198*, 177–192.

Kuttykrishnan, S., Sabina, J., Langton, L. L., Johnston, M., & Brent, M. R. (2010). A quantitative model of glucose signaling in yeast reveals an incoherent feed forward loop leading to a specific, transient pulse of transcription. *Proceedings of the National Academy of Sciences of the United States of America*, *107*, 16743–16748.

Lagunas, R., & Gancedo, J. M. (1973). Reduced pyridine-nucleotides balance in glucose-growing Saccharomyces cerevisiae. *European Journal of Biochemistry*, *37*, 90–94.

Lauff, D. B., & Santa-Maria, G. E. (2010). Potassium deprivation is sufficient to induce a cell death program in Saccharomyces cerevisiae. *FEMS Yeast Research*, *10*, 497–507.

Lewis, N. E., Nagarajan, H., & Palsson, B. O. (2012). Constraining the metabolic genotype-phenotype relationship using a phylogeny of in silico methods. *Nature Reviews Microbiology*, *10*, 291–305.

Li, W., & Mitchell, A. P. (1997). Proteolytic activation of Rim1p, a positive regulator of yeast sporulation and invasive growth. *Genetics*, *145*, 63–73.

Loukin, S. H., Vaillant, B., Zhou, X. L., Spalding, E. P., Kung, C., & Saimi, Y. (1997). Random mutagenesis reveals a region important for gating of the yeast K^+ channel Ykc1. *The EMBO Journal*, *16*, 4817–4825.

Lubin, M., & Ennis, H. (1964). On the role of intracellular potassium in protein synthesis. *Biochimica et Biophysica Acta*, *80*, 614–631.

Madrid, R., Gomez, M. J., Ramos, J., & Rodriguez-Navarro, A. (1998). Ectopic potassium uptake in trk1 trk2 mutants of Saccharomyces cerevisiae correlates with a highly hyperpolarized membrane potential. *The Journal of Biological Chemistry*, *273*, 14838–14844.

Maeta, K., Izawa, S., Okazaki, S., Kuge, S., & Inoue, Y. (2004). Activity of the Yap1 transcription factor in Saccharomyces cerevisiae is modulated by methylglyoxal, a metabolite derived from glycolysis. *Molecular and Cellular Biology*, *24*, 8753–8764.

Maresova, L., & Sychrova, H. (2005). Physiological characterization of Saccharomyces cerevisiae kha1 deletion mutants. *Molecular Microbiology*, *55*, 588–600.

Maresova, L., Urbankova, E., Gaskova, D., & Sychrova, H. (2006). Measurements of plasma membrane potential changes in Saccharomyces cerevisiae cells reveal the importance of the Tok1 channel in membrane potential maintenance. *FEMS Yeast Research, 6*, 1039–1046.

Maresova, L., Vydareny, T., & Sychrova, H. (2012). Comparison of the influence of small GTPases Arl1 and Ypt6 on yeast cells' tolerance to various stress factors. *FEMS Yeast Research, 12*, 332–340.

Marquina, M., Gonzalez, A., Barreto, L., Gelis, S., Munoz, I., Ruiz, A., et al. (2012). Modulation of yeast alkaline cation tolerance by Ypi1 requires calcineurin. *Genetics, 190*, 1355–1364.

Mendizabal, I., Pascual-Ahuir, A., Serrano, R., & de Larrinoa, I. F. (2001). Promoter sequences regulated by the calcineurin-activated transcription factor Crz1 in the yeast ENA1 gene. *Molecular Genetics and Genomics, 265*, 801–811.

Mendoza, I., Rubio, F., Rodriguez-Navarro, A., & Pardo, J. M. (1994). The protein phosphatase calcineurin is essential for NaCl tolerance of *Saccharomyces cerevisiae*. *The Journal of Biological Chemistry, 269*, 8792–8796.

Merchan, S., Bernal, D., Serrano, R., & Yenush, L. (2004). Response of the *Saccharomyces cerevisiae* Mpk1 mitogen-activated protein kinase pathway to increases in internal turgor pressure caused by loss of Ppz protein phosphatases. *Eukaryotic Cell, 3*, 100–107.

Mo, M. L., Palsson, B. O., & Herrgard, M. J. (2009). Connecting extracellular metabolomic measurements to intracellular flux states in yeast. *BMC Systems Biology, 3*, 37.

Mulet, J. M., Leube, M. P., Kron, S. J., Rios, G., Fink, G. R., & Serrano, R. (1999). A novel mechanism of ion homeostasis and salt tolerance in yeast: The Hal4 and Hal5 protein kinases modulate the Trk1-Trk2 potassium transporter. *Molecular and Cellular Biology, 19*, 3328–3337.

Mulet, J. M., Llopis-Torregrosa, V., Primo, C., Marques, M. C., & Yenush, L. (2013). Endocytic regulation of alkali metal transport proteins in mammals, yeast and plants. *Current Genetics, 59*, 207–230.

Munson, A. M., Haydon, D. H., Love, S. L., Fell, G. L., Palanivel, V. R., & Rosenwald, A. G. (2004). Yeast ARL1 encodes a regulator of K^+ influx. *Journal of Cell Science, 117*, 2309–2320.

Nass, R., & Rao, R. (1998). Novel localization of a Na^+/H^+ exchanger in a late endosomal compartment of yeast. Implications for vacuole biogenesis. *The Journal of Biological Chemistry, 273*, 21054–21060.

Navarrete, C., Petrezselyova, S., Barreto, L., Martinez, J. L., Zahradka, J., Arino, J., et al. (2010). Lack of main K^+ uptake systems in Saccharomyces cerevisiae cells affects yeast performance in both potassium-sufficient and potassium-limiting conditions. *FEMS Yeast Research, 10*, 508–517.

Ohgaki, R., Nakamura, N., Mitsui, K., & Kanazawa, H. (2005). Characterization of the ion transport activity of the budding yeast Na^+/H^+ antiporter, Nha1p, using isolated secretory vesicles. *Biochimica et Biophysica Acta, 1712*, 185–196.

Okorokov, L. A., Lichko, L. P., & Kulaev, I. S. (1980). Vacuoles: Main compartments of potassium, magnesium, and phosphate ions in Saccharomyces carlsbergenis cells. *Journal of Bacteriology, 144*, 661–665.

Orij, R., Brul, S., & Smits, G. J. (2011). Intracellular pH is a tightly controlled signal in yeast. *Biochimica et Biophysica Acta, 1810*, 933–944.

Page, M. J., & Di Cera, E. (2006). Role of Na^+ and K^+ in enzyme function. *Physiological Reviews, 86*, 1049–1092.

Palmer, C. P., Zhou, X. L., Lin, J., Loukin, S. H., Kung, C., & Saimi, Y. (2001). A TRP homolog in Saccharomyces cerevisiae forms an intracellular Ca(2+)-permeable channel in the yeast vacuolar membrane. *Proceedings of the National Academy of Sciences of the United States of America, 98*, 7801–7805.

Parrou, J. L., Teste, M. A., & Francois, J. (1997). Effects of various types of stress on the metabolism of reserve carbohydrates in Saccharomyces cerevisiae: Genetic evidence for a stress-induced recycling of glycogen and trehalose. *Microbiology, 143*(Pt 6), 1891–1900.

Pedelini, L., Marquina, M., Arino, J., Casamayor, A., Sanz, L., Bollen, M., et al. (2007). YPI1 and SDS22 proteins regulate the nuclear localization and function of yeast type 1 phosphatase Glc7. *The Journal of Biological Chemistry, 282*, 3282–3292.

Perez-Valle, J., Jenkins, H., Merchan, S., Montiel, V., Ramos, J., Sharma, S., et al. (2007). Key role for intracellular K^+ and protein kinases Sat4/Hal4 and Hal5 in the plasma membrane stabilization of yeast nutrient transporters. *Molecular and Cellular Biology, 27*, 5725–5736.

Perez-Valle, J., Rothe, J., Primo, C., Martinez, P. M., Arino, J., Pascual-Ahuir, A., et al. (2010). Hal4 and Hal5 protein kinases are required for general control of carbon and nitrogen uptake and metabolism. *Eukaryotic Cell, 9*, 1881–1890.

Petrezselyova, S., Kinclova-Zimmermannova, O., & Sychrova, H. (2013). Vhc1, a novel transporter belonging to the family of electroneutral cation-Cl(−) cotransporters, participates in the regulation of cation content and morphology of Saccharomyces cerevisiae vacuoles. *Biochimica et Biophysica Acta, 1828*, 623–631.

Petrezselyova, S., Ramos, J., & Sychrova, H. (2011). Trk2 transporter is a relevant player in K^+ supply and plasma-membrane potential control in Saccharomyces cerevisiae. *Folia Microbiologica, 56*, 23–28.

Petrezselyova, S., Zahradka, J., & Sychrova, H. (2010). Saccharomyces cerevisiae BY4741 and W303-1A laboratory strains differ in salt tolerance. *Fungal Biology, 114*, 144–150.

Plášek, J., Gášková, D., Lichtenberg-Fraté, H., Ludwig, J., & Höfer, M. (2012). Monitoring of real changes of plasma membrane potential by diS-C(3)(3) fluorescence in yeast cell suspensions. *Journal of Bioenergetics and Biomembranes, 44*, 559–569.

Plášek, J., Gášková, D., Ludwig, J., & Höfer, M. (2013). Early changes in membrane potential of Saccharomyces cerevisiae induced by varying extracellular K^+, Na^+ or H^+. *Journal of Bioenergetics and Biomembranes, 45*, 561–568.

Proft, M., & Struhl, K. (2004). MAP kinase-mediated stress relief that precedes and regulates the timing of transcriptional induction. *Cell, 118*, 351–361.

Raue, A., Kreutz, C., Maiwald, T., Bachmann, J., Schilling, M., Klingmuller, U., et al. (2009). Structural and practical identifiability analysis of partially observed dynamical models by exploiting the profile likelihood. *Bioinformatics, 25*, 1923–1929.

Rivetta, A., Kuroda, T., & Slayman, C. (2011). Anion currents in yeast K^+ transporters (TRK) characterize a structural homologue of ligand-gated ion channels. *Pflügers Archiv, 462*, 315–330.

Rocca-Serra, P., Brandizi, M., Maguire, E., Sklyar, N., Taylor, C., Begley, K., et al. (2010). ISA software suite: Supporting standards-compliant experimental annotation and enabling curation at the community level. *Bioinformatics, 26*, 2354–2356.

Rodriguez-Navarro, A. (2000). Potassium transport in fungi and plants. *Biochimica et Biophysica Acta, 1469*, 1–30.

Rodriguez-Navarro, A., & Ramos, J. (1984). Dual system for potassium transport in *Saccharomyces cerevisiae*. *Journal of Bacteriology, 159*, 940–945.

Roomans, G. M., & Seveus, L. A. (1976). Subcellular localization of diffusible ions in the yeast Saccharomyces cerevisiae: Quantitative microprobe analysis of thin freeze-dried sections. *Journal of Cell Science, 21*, 119–127.

Rosenfeld, L., Reddi, A. R., Leung, E., Aranda, K., Jensen, L. T., & Culotta, V. C. (2010). The effect of phosphate accumulation on metal ion homeostasis in Saccharomyces cerevisiae. *Journal of Biological Inorganic Chemistry, 15*, 1051–1062.

Ruiz, A., & Arino, J. (2007). Function and regulation of the Saccharomyces cerevisiae ENA sodium ATPase system. *Eukaryotic Cell, 6*, 2175–2183.

Ruiz, A., Gonzalez, A., Munoz, I., Serrano, R., Abrie, J. A., Strauss, E., et al. (2009). Moonlighting proteins Hal3 and Vhs3 form a heteromeric PPCDC with Ykl088w in yeast CoA biosynthesis. *Nature Chemical Biology*, *5*, 920–928.

Ruiz, A., Munoz, I., Serrano, R., Gonzalez, A., Simon, E., & Arino, J. (2004). Functional characterization of the *Saccharomyces cerevisiae* VHS3 gene: A regulatory subunit of the Ppz1 protein phosphatase with novel, phosphatase-unrelated functions. *The Journal of Biological Chemistry*, *279*, 34421–34430.

Ruiz, A., Serrano, R., & Arino, J. (2008). Direct regulation of genes involved in glucose utilization by the calcium/calcineurin pathway. *The Journal of Biological Chemistry*, *283*, 13923–13933.

Schaber, J., Baltanas, R., Bush, A., Klipp, E., & Colman-Lerner, A. (2012). Modelling reveals novel roles of two parallel signalling pathways and homeostatic feedbacks in yeast. *Molecular Systems Biology*, *8*, 622.

Serrano, R. (1983). In vivo glucose activation of the yeast plasma membrane ATPase. *FEBS Letters*, *156*, 11–14.

Stefan, C. P., Zhang, N., Sokabe, T., Rivetta, A., Slayman, C. L., Montell, C., et al. (2013). Activation of an essential calcium signaling pathway in Saccharomyces cerevisiae by Kch1 and Kch2, putative low-affinity potassium transporters. *Eukaryotic Cell*, *12*, 204–214.

Taylor, C. F., Field, D., Sansone, S. A., Aerts, J., Apweiler, R., Ashburner, M., et al. (2008). Promoting coherent minimum reporting guidelines for biological and biomedical investigations: The MIBBI project. *Nature Biotechnology*, *26*, 889–896.

Terpitz, U., Raimunda, D., Westhoff, M., Sukhorukov, V. L., Beauge, L., Bamberg, E., et al. (2008). Electrofused giant protoplasts of Saccharomyces cerevisiae as a novel system for electrophysiological studies on membrane proteins. *Biochimica et Biophysica Acta*, *1778*, 1493–1500.

Tiger, C. F., Krause, F., Cedersund, G., Palmer, R., Klipp, E., Hohmann, S., et al. (2012). A framework for mapping, visualisation and automatic model creation of signal-transduction networks. *Molecular Systems Biology*, *8*, 578.

van Heusden, G. P. (2009). 14-3-3 Proteins: Insights from genome-wide studies in yeast. *Genomics*, *94*, 287–293.

Venturi, G. M., Bloecher, A., Williams-Hart, T., & Tatchell, K. (2000). Genetic interactions between GLC7, PPZ1 and PPZ2 in saccharomyces cerevisiae. *Genetics*, *155*, 69–83.

Viladevall, L., Serrano, R., Ruiz, A., Domenech, G., Giraldo, J., Barcelo, A., et al. (2004). Characterization of the calcium-mediated response to alkaline stress in *Saccharomyces cerevisiae*. *The Journal of Biological Chemistry*, *279*, 43614–43624.

Weiss, T. F. (1996). *Cellular biophysics—Transport*. Cambridge: Bradford Book.

Wieland, J., Nitsche, A. M., Strayle, J., Steiner, H., & Rudolph, H. K. (1995). The PMR2 gene cluster encodes functionally distinct isoforms of a putative Na^+ pump in the yeast plasma membrane. *The EMBO Journal*, *14*, 3870–3882.

Williams-Hart, T., Wu, X., & Tatchell, K. (2002). Protein phosphatase type 1 regulates ion homeostasis in Saccharomyces cerevisiae. *Genetics*, *160*, 1423–1437.

Wolstencroft, K., Owen, S., Horridge, M., Krebs, O., Mueller, W., Snoep, J. L., et al. (2011). RightField: Embedding ontology annotation in spreadsheets. *Bioinformatics*, *27*, 2021–2022.

Yenush, L., Merchan, S., Holmes, J., & Serrano, R. (2005). pH-Responsive, posttranslational regulation of the Trk1 potassium transporter by the type 1-related Ppz1 phosphatase. *Molecular and Cellular Biology*, *25*, 8683–8692.

Yenush, L., Mulet, J. M., Arino, J., & Serrano, R. (2002). The Ppz protein phosphatases are key regulators of K^+ and pH homeostasis: Implications for salt tolerance, cell wall integrity and cell cycle progression. *The EMBO Journal*, *21*, 920–929.

Yoshimoto, H., Saltsman, K., Gasch, A. P., Li, H. X., Ogawa, N., Botstein, D., et al. (2002). Genome-wide analysis of gene expression regulated by the calcineurin/Crz1p signaling pathway in *Saccharomyces cerevisiae*. *The Journal of Biological Chemistry*, *277*, 31079–31088.

Zahradka, J., & Sychrova, H. (2012). Plasma-membrane hyperpolarization diminishes the cation efflux via Nha1 antiporter and Ena ATPase under potassium-limiting conditions. *FEMS Yeast Research*, *12*, 439–446.

Zahradka, J., van Heusden, G. P., & Sychrova, H. (2012). Yeast 14-3-3 proteins participate in the regulation of cell cation homeostasis via interaction with Nha1 alkali-metal-cation/proton antiporter. *Biochimica et Biophysica Acta*, *1820*, 849–858.

Zhou, X. L., Loukin, S. H., Coria, R., Kung, C., & Saimi, Y. (2005). Heterologously expressed fungal transient receptor potential channels retain mechanosensitivity in vitro and osmotic response in vivo. *European Biophysics Journal*, *34*, 413–422.

Zi, Z. (2011). SBML-PET-MPI: A parallel parameter estimation tool for Systems Biology Markup Language based models. *Bioinformatics*, *27*, 1028–1029.

Zotova, L., Aleschko, M., Sponder, G., Baumgartner, R., Reipert, S., Prinz, M., et al. (2010). Novel components of an active mitochondrial K(+)/H(+) exchange. *The Journal of Biological Chemistry*, *285*, 14399–14414.

CHAPTER TWO

Towards a Systems Level Understanding of the Oxygen Response of *Escherichia coli*

Katja Bettenbrock*[,1], Hao Bai[†], Michael Ederer[‡], Jeffrey Green[§], Klaas J. Hellingwerf[¶], Michael Holcombe[†], Samantha Kunz[‡], Matthew D. Rolfe[§], Guido Sanguinetti[#], Oliver Sawodny[‡], Poonam Sharma[§], Sonja Steinsiek*, Robert K. Poole[§]

*Max Planck Institute for Dynamics of Complex Technical Systems, Magdeburg, Germany
[†]Department of Computer Science, The University of Sheffield, Sheffield, United Kingdom
[‡]Institute for System Dynamics, University of Stuttgart, Stuttgart, Germany
[§]Department of Molecular Biology and Biotechnology, The University of Sheffield, Sheffield, United Kingdom
[¶]Swammerdam Institute for Life Sciences, University of Amsterdam, Amsterdam, The Netherlands
[#]School of Informatics, University of Edinburgh, Edinburgh, United Kingdom
[1]Corresponding author: e-mail address: bettenbrock@mpi-magdeburg.mpg.de

Contents

Advances in Microbial Physiology, Volume 64
ISSN 0065-2911
http://dx.doi.org/10.1016/B978-0-12-800143-1.00002-6

Abstract

Escherichia coli is a facultatively anaerobic bacterium. With glucose if no external electron acceptors are available, ATP is produced by substrate level phosphorylation. The intracellular redox balance is maintained by mixed-acid fermentation, that is, the production and excretion of several organic acids. When oxygen is available, *E. coli* switches to aerobic respiration to achieve redox balance and optimal energy conservation by proton translocation linked to electron transfer. The switch between fermentative and aerobic respiratory growth is driven by extensive changes in gene expression and protein synthesis, resulting in global changes in metabolic fluxes and metabolite concentrations. This oxygen response is determined by the interaction of global and local genetic regulatory mechanisms, as well as by enzymatic regulation. The response is affected by basic physical constraints such as diffusion, thermodynamics and the requirement for a balance of carbon, electrons and energy (predominantly the proton motive force and the ATP pool). A comprehensive systems level understanding of the oxygen response of *E. coli* requires the integrated interpretation of experimental data that are pertinent to the multiple levels of organization that mediate the response. In the pan-European venture, *Systems Biology of Microorganisms* (SysMO) and specifically within the project Systems Understanding of Microbial Oxygen Metabolism (SUMO), regulator activities, gene expression, metabolite levels and metabolic flux datasets were obtained using a standardized and reproducible chemostat-based experimental system. These different types and qualities of data were integrated using mathematical models. The approach described here has revealed a much more detailed picture of the aerobic–anaerobic response, especially for the environmentally critical microaerobic range that is located between unlimited oxygen availability and anaerobiosis.

ABBREVIATIONS

ArcBA aerobic respiratory control two-component system
AU acetate units
ETC electron transport chain
FNR fumarate nitrate reduction regulator
FRD fumarate reductase
PAS Per-Arnt-Sim domain
PMF proton motive force
PTS phosphotransferase system
SDH succinate dehydrogenase
TCA tricarboxylic acid
TF transcription factor
TKM thermodynamic-kinetic modelling
UPEC uropathogenic *E. coli*

1. INTRODUCTION

Almost all bacteria exhibit complex networks of respiratory pathways quite unlike the near–linear routes of electron transfer to oxygen extant in almost all mitochondria (Nicholls & Ferguson, 2013). Exceptions in the eukaryotic domain do exist, as in the presence of alternative, cyanide-resistant oxidases that occur in certain plants and fungi (Moore et al., 2008). When such alternatives exist, the function and rationale of the branching is often understood or can be assumed. Oxidases that are 'uncoupled' from energy generation are thought to be important in thermogenesis or spilling reactions in which excess reductant or oxidant can be 'burnt' (Wagner, Krab, Wagner, & Moore, 2008). Nevertheless, most routes of electron transfer to oxygen in mitochondria are linear—although immensely complex in terms of the subunit compositions of the individual respiratory complexes. For a recent account, see Nicholls and Ferguson (2013).

Such minimally branched or unbranched pathways appear to be extremely rare in bacteria. Furthermore, the functions of the multiple electron transport chains (ETCs) in bacteria are often poorly understood leading to the suggestion that some pathways may be redundant. A more likely interpretation is that the subtleties of electron transport and energetic functions that have led to the acquisition and persistence of such complexity during evolution is not yet understood. Examples of branched respiratory systems in bacteria are reviewed by Poole and Cook (2000).

In comparison to most other organisms, the biochemistry, physiology and molecular biology of the Gram-negative enterobacterium *Escherichia coli* are much better understood. It is therefore astonishing that, even in this paradigm amongst laboratory-evolved bacteria (as is the case for *E. coli* K-12, derived after curing of the F plasmid from a 1922 isolate from a convalescent diphtheria patient), we still do not really understand why the pathways of electron and hydrogen transfer from low-redox potential 'donors' to high-potential acceptors (oxygen, nitrate, fumarate and numerous others) are so convoluted. This review deals only with oxygen as the pre-eminent electron acceptor but similar questions could be posed, and perhaps answered, by equivalent systems approaches, for electron transfer to nitrate or nitrite, for example, and such experiments and modelling studies are under way in the authors' laboratories.

2. THE METABOLIC MODES OF *E. COLI* AND OXYGEN LIMITATION

E. coli possesses a versatile metabolic repertoire that is well adapted to promote survival and reproduction in the gastrointestinal tract of a suitable host (Ingledew & Poole, 1984) and, in the case of certain pathogenic strains such as uropathogenic *E. coli* (UPECs), in other tissues and locations (Slonczewski & Foster, 2008). *E. coli* can thrive on a relatively limited number of low-molecular mass compounds, sometimes available only transiently and at low concentrations (for references, see Ingledew & Poole, 1984). Success in the scavenging lifestyle is made possible by the presence of high affinity or high-capacity transport systems that are powered directly by ATP or ultimately by membrane-bound oxido-reduction reactions. These generate a transmembrane electrochemical gradient, predominantly a proton motive force (PMF), although Na^+ gradients play lesser roles, which can either drive the transport system directly or via the formation of ATP by F_1F_o ATP synthase.

The redox energy afforded by substrate oxidation is conserved by critical oxido-reduction systems that require one or more terminal electron acceptors of higher redox potential than the donor. Pre-eminent amongst these electron acceptor reactions, at least when a trace of molecular oxygen is available, is the four-electron reduction of oxygen to water. In the absence of oxygen, *E. coli* can utilize a range of alternative electron acceptors such as nitrate, fumarate, trimethylamine-*N*-oxide and others. Without molecular oxygen or other electron acceptors, mixed-acid fermentation is used for the regeneration of the NAD^+ pool. Thus, *E. coli* operates a metabolic hierarchy in which aerobic respiration is preferred to anaerobic respiration, which in turn is preferred to fermentation. This hierarchy reflects the relative amounts of energy that can be conserved by each metabolic mode and is maintained by the coordinated action of transcription factors (TFs) that reprogramme gene expression in response to key environmental signals such as oxygen, nitrate and nitrite.

The bacterium *E. coli* has long been a model-organism for studying these metabolic modes (viz., aerobic respiration, anaerobic respiration and fermentation). However, much less well characterized are the states that exist between aerobiosis with plentiful oxygen and anaerobiosis without any oxygen, often called the microaerobic range, where molecular oxygen is present but at concentrations too low to allow for a complete aerobic respiratory lifestyle. Although microaerobic environments are frequently encountered by *E. coli* and many other bacteria, these conditions are less well studied due largely to the fact that it is difficult to define reproducible microaerobic conditions. The Systems Understanding of Microbial Oxygen Metabolism (SUMO) consortium was supported by the ERA SysMO initiative (http://www.sysmo.net/) to obtain a systems level understanding of the responses of *E. coli* towards oxygen including the microaerobic range. The experimental studies in the different partner laboratories were performed using the same well-defined strain, namely *E. coli* K-12 MG1655, grown in a common medium in chemostats in which oxygen provision was controlled. Because important physiological adaptations occur at oxygen tensions below the sensitivity of conventional membrane-covered Clark oxygen electrodes, a method of oxygen monitoring (the aerobiosis scale based on the acetate unit, AU) that is based on the physiological consequences of oxygen limitation rather than the detection of unused oxygen in the culture was adopted. The experimental studies were complemented by modelling approaches covering different levels of complexity. The outcomes of the SUMO consortium greatly expanded the knowledge concerning physiology and regulation of *E. coli* at microaerobic conditions.

2.1. The aerobiosis scale

As noted above, *E. coli* has three basic metabolic modes. However, when grown in media that contain a large excess of a carbon source like glucose, *E. coli* will, despite the presence of abundant oxygen, convert a large fraction of the incoming sugar into acetate, via the acetate kinase (*ackA*)/phospho-transacetylase (*pta*) pathway (Brown, Jones-Mortimer, & Kornberg, 1977; Clark, 1989; Landwall & Holme, 1977). If this pathway is disrupted by mutation, as in a *pta* deletion strain, the rate of acetate production is strongly decreased without serious implications for growth rate. This aerobic 'overflow' metabolism is explained in the literature as being due to an overcapacity of the enzymatic machinery of the upper part of the glycolytic pathway, as compared to the lower part (El-Mansi & Holms, 1989), or to an imbalance between the rate of glucose uptake and the (carbon) demands for both biosynthesis and energy production because of either a limiting capacity of the tricarboxylic acid (TCA) cycle or limited respiratory chain capacity (El-Mansi & Holms, 1989; Holms, 1996). In agreement with this, it is observed that when the glucose-influx capacity is reduced, for example, by deletion of (a) key component(s) of the glucose phosphotransferase system (PTS) (Picon, Teixeira de Mattos, & Postma, 2005), this acetate production is almost completely abolished.

The practical outcome of these metabolic adaptations is that, whilst oxygen is sufficient in a glucose-limited culture, acetate flux from the cell is negligible. As oxygen availability falls, there is a near–linear relationship between the acetate detectable in culture supernatants and the rate of oxygen supply (Alexeeva, Hellingwerf, & Teixeira de Mattos, 2002). Thus, in this approach, one measures the physiological consequences of oxygen limitation rather than the detection of unused oxygen in the culture. By calibrating each chemostat under each condition in each laboratory, we can define oxygen provision (rate). In the SUMO consortium's work, the minimum oxygen concentration at which zero acetate is detectable is defined as 100% aerobiosis (100% AU). As oxygen provision decreases (by supplying less oxygen into the gas flow under otherwise identical conditions), there is a progressive increase in acetate flux that reaches a maximum under anoxic conditions. Thus aerobiosis is given as a % value below 100 to signify oxygen limitation, or above 100% to signify a degree of oxygen provision that exceeds the minimum needed to maintain zero acetate flux. An example is given in Rolfe et al. (2011). By calibrating each chemostat, even from different manufacturers and in different laboratories it is possible to perform geographically isolated but strictly comparable experiments and combine the data for analysis and modelling. A clear output from this work is that measurement of residual oxygen in the culture, for example using an oxygen electrode, is not adequate in defining the microaerobic states in which important physiological changes occur. Similarly, measurement of oxygen diffusion rates into the culture (K_La), whilst useful for assessing the efficiency of different aeration methods and chemostat designs (Pirt, 1985), cannot give information on the ability of that oxygen flux to maintain a fully aerobic physiology.

2.2. Oxygen-sensing TFs in *E. coli*

The profound physiological changes that occur in response to oxygen availability are indicative of the imperative for *E. coli* to monitor oxygen. There are two major oxygen-responsive regulators in *E. coli* that together coordinate the reprogramming of gene expression to ensure that the optimal metabolic mode is adopted, that is, fumarate nitrate reduction regulator (FNR) and ArcA. The FNR protein is member of the CRP-FNR superfamily of TFs, characterized by the presence of an N-terminal sensory domain and a

C-terminal DNA-binding domain (Green, Scott, & Guest, 2001). The FNR protein is able to regulate transcription of target genes under anaerobic conditions by incorporation of an oxygen-labile [4Fe-4S] cluster into the sensory domain of the protein. Iron–sulphur cluster acquisition results in the formation of FNR dimers from two FNR monomers. The FNR dimer has enhanced DNA-binding activity and binds in a sequence-specific manner to target promoters, where it acts as an activator of anaerobic genes and a repressor of aerobic genes. In the presence of oxygen, the [4Fe–4S] cluster is degraded to a [2Fe–2S] cluster via a [3Fe–4S] intermediate. As a consequence, the FNR dimers breakdown and site-specific DNA-binding is inhibited. The [2Fe–2S] form of FNR stores sulphide, released from the breakdown of the [4Fe-4S] cluster facilitating facile repair of the iron–sulphur cluster and restoration of FNR activity during periods of fluctuating oxygen availability (Zhang et al., 2012). However if oxygen persists in the environment, the [2Fe–2S] form of FNR degrades to leave apoFNR, which can acquire a [4Fe–4S] cluster by interaction with the cell's iron–sulphur cluster biosynthetic machinery. The details of oxygen sensing by FNR have been recently reviewed (Crack, Green, Hutchings, Thomson, & Le Brun, 2012).

In contrast to FNR, the aerobic respiratory control two-component system (ArcBA) is an indirect sensor of oxygen. This system is comprised of the ArcB protein, a tripartite membrane-bound sensor kinase, and ArcA, the cognate response regulator. ArcB belongs to the subclass of hybrid sensor kinases that in *E. coli* also includes BarA, BvgS, EvgS, LemA (GacS), RteA and TorS (Alvarez & Georgellis, 2010; Malpica, Sandoval, Rodríguez, Franco, & Georgellis, 2006). The tripartite kinase ArcB possesses a transmembrane domain comprising 16-amino acid residues, followed by a leucine-zipper- and a Per-Arnt-Sim (PAS) domain that connects the transmembrane domain to the catalytic domains of ArcB. PAS domains are versatile sensory structures that are associated with monitoring changes in light, redox potential and the overall energy state of cells (Taylor & Zhulin, 1999). The cytosolic PAS domain of each ArcB monomer contains two key cysteine residues (Cys180 and Cys241) that must be in the reduced (thiol) state for maximal ArcB autokinase activity. The oxidation state of the ArcB sensory Cys residues is set by monitoring the redox state of quinones in the ETC, with fermentation products acting as allosteric effectors to accelerate ArcB kinase activity (Georgellis, Kwon, & Lin, 2001; Malpica, Franco, Rodriguez, Kwon, & Georgellis, 2004; Rodriguez, Kwon, & Georgellis, 2004). When oxygen becomes available, as occurs with increasing aerobiosis

(Alexeeva, Hellingwerf, & Teixeira de Mattos, 2003), ArcB activity gradually shifts from autokinase activity to phosphatase activity via the formation of two intermolecular disulphide bonds resulting in the formation of a covalently linked ArcB dimer. In the reduced state, ArcB undergoes autophosphorylation of His292 in the primary transmitter domain (H1) of ArcB, the phosphate group is passed to Asp576 in the ArcB receiver domain (D1) and then to His717 within the phosphotransfer domain (H2) (Georgellis, Lynch, & Lin, 1997; Iuchi & Lin, 1992a, 1992b). The phosphotransfer is completed when ArcB interacts with ArcA to phosphorylate a conserved Asp residue (Asp54) in the N-terminal receiver domain of ArcA. ArcA is a typical response regulator with a C-terminal helix–turn–helix DNA-binding domain. Upon phosphorylation of Asp54, ArcA becomes an octameric sequence-specific DNA-binding protein (Jeon, Lee, Han, Kim, & Hwang, 2001), allowing it to control the expression of multiple operons across the *E. coli* chromosome (Liu & De Wulf, 2004; Salmon et al., 2005). The system is switched off by the phosphatase activity of the oxidized (disulphide) form of ArcB leading to the dephosphorylation of ArcA and inhibition of DNA binding (Georgellis, 1998).

2.3. Oxygen-utilizing systems in *E. coli*

A comprehensive consideration of the oxygen metabolism of *E. coli* clearly requires an understanding of the nature of the oxygen-reactive proteins and systems in this organism. The contributing oxygen-consuming proteins are terminal oxidases, one haemoglobin and various oxygenases, such as those involved in quinone biosynthesis (Søballe & Poole, 1999) and diverse biosynthetic modules. The authors are unaware of calculations of the relative contribution of these in *E. coli*, but it is generally assumed that oxygen consumption in the biosphere is >90% attributable to terminal oxidases that reduce oxygen to water (Wikström, Krab, & Saraste, 1981). The haemoglobin of *E. coli* is the flavohaemoglobin, Hmp whose only well-defined function is NO detoxification (Poole & Hughes, 2000) when it consumes 1O_2 per NO to generate nitrate. In normal growth conditions, in the absence of NO, the activity of this protein is minimal, although anaerobic cultures supplemented not with NO, but with nitrate or nitrite show very high levels of expression of this globin (Poole et al., 1996). Furthermore, the affinity for oxygen is low (Mills, Sedelnikova, Søballe, Hughes, & Poole, 2001) so that in the natural environment of the gut, where oxygen would be below the K_m of Hmp, its contribution can be ignored. Here, the focus is on the three genetically and (partially) biochemically defined oxidases.

2.3.1 *Cytochrome* bo′

The older literature described two terminal oxidases in *E. coli*; these are cytochrome *bd* (formerly a_2) and cytochrome *bo′* (often called, but inappropriately in the view of some, bo_3) (Poole & Chance, 1995). Cytochrome *bo′* is a member of the pre-eminent haem-Cu superfamily of oxidases that contain, at the active site where oxygen is bound and reduced, a high-spin haem (haem O in *E. coli*, haem a_3 in mitochondria) and, in close proximity, a magnetically coupled Cu atom (Cu_B). Together these cofactors constitute a binuclear active site where oxygen (and competing ligands such as CO and NO) bind and may be reduced. The *E. coli* cytochrome *bo′*; however, retains only one of the pair of Cu atoms that are present in the paradigm of this class of oxidases, mitochondrial oxidase cytochrome aa_3 (Nicholls & Ferguson, 2013; Wikström et al., 1981). The Cu_A centre, whose function is to shuttle electrons from the immediate donor, cytochrome *c*, to the binuclear centre in cytochrome aa_3 is absent in *E. coli* cytochrome *bo′*. This is because in the latter the reductant is not cytochrome *c* but ubiquinol; hence, a quinol-binding site functionally replaces Cu_A. *E. coli* cytochrome *bo′* is encoded by the *cyoABCDE* operon; the first four genes encode four hydrophobic protein subunits of the functional oxidase, whilst *cyoE* encodes a protohaem IX farnesyltransferase that is involved in the synthesis of the haem O prosthetic group (Anraku, 1988; Puustinen, Finel, Haltia, Gennis, & Wikström, 1991). The crystal structure of *E. coli* cytochrome *bo′* has been solved (Abramson, Larsson, et al., 2000; Abramson, Riistama, et al., 2000) and led to important conclusions about structure-function relationships that are outside the scope of this review. Some details of the assembly process are available but not pertinent here (Du Plessis, Nouwen, & Driessen, 2006; Saiki, Nakamura, Mogi, & Anraku, 1996; Stenberg, von Heijne, & Daley, 2007).

Cytochrome *bo′* catalyses the four-electron reduction of molecular oxygen to water using ubiquinol as electron donor. It functions as a proton pump, conserving energy from quinol oxidation as a PMF ($H^+/O=2$) (Puustinen, Finel, Virkki, & Wikström, 1989). The oxygen kinetics of this oxidase as well as of the two cytochrome *bd*-type oxidases is described below.

2.3.2 *Cytochrome* bd-I

Cytochrome *bd-I* is a terminal respiratory oxidase that has been found only in prokaryotes; it is functionally and structurally distinct from cytochrome *bo′* and all other oxidases, but there is no crystal structure of any oxidase in this class. Cytochrome *bd* couples electron transfer from quinol to molecular oxygen (reducing the latter to water) with generation of an electrochemical proton gradient across the cytoplasmic membrane; for a review see, Borisov, Gennis, Hemp, and Verkhovsky (2011); however, the energetic efficiency of coupling is twofold lower than that of the cytochrome *bo′* quinol oxidase and aa_3-type cytochrome oxidases (Calhoun, Oden, Gennis, de Mattos, & Neijssel, 1993). Cytochrome *bd* oxidase is usually considered to prevail under low-oxygen conditions (for a review, see Poole & Cook, 2000) but the presence in media of cyanide, NO (Mason et al., 2009) or CO (Davidge et al., 2009) also induce its expression. It is observed *in vivo* and in membranes as a stable oxygenated complex (Poole, Salmon, & Chance, 1983) and is also generally isolated in this form (Belevich, Borisov, Konstantinov, & Verkhovsky, 2005), in agreement with its high affinity for oxygen during turnover (see below).

In contrast to the class of haem-copper oxidases, whose various members have been investigated in great detail, cytochrome *bd* remains poorly understood at the structural level. The integral membrane protein complex is composed of two different subunits carrying three haems, b_{558}, b_{595} and *d*, which are probably located near the periplasmic side of the membrane (Zhang, Barquera, & Gennis, 2004). This may have relevance to the following discussion on oxygen access to its site(s) of reduction but the ligand is generally considered to be highly mobile across phospholipid bilayers (Arras, Schirawski, & Unden, 1998). The low-spin hexacoordinate haem b_{558} seems to be directly involved in quinol oxidation, and the two quinol protons are released into the periplasm (Mogi et al., 2006). His_{186} and Met_{393} of subunit I are identified as its axial ligands (Kaysser, Ghaim, Georgiou, & Gennis, 1995). The high-spin haem *d* is the core of the enzyme where oxygen is bound (Poole et al., 1983), activated and reduced into H_2O. The protons required to reduce oxygen are probably taken from the cytoplasmic side of the membrane via an extended transmembrane H^+-transfer pathway, as in the typical aa_3-type oxidases (Borisov, Belevich, Bloch, Mogi, & Verkhovsky, 2008). Thus, charge separation results in an electrochemical proton gradient across the membrane but no direct proton pumping (as in the haem-Cu superfamily occurs). The oxidase *in vivo* is thought to undergo the following catalytic transformations:

R → oxy → P → F → R, where R, oxy and F are, respectively, reduced, oxygenated and ferryl forms of haem *d* (Borisov, Gennis, et al., 2011).

A key issue in studies of the functioning of *bd*-type terminal oxidases, and the present review, is to understand how the oxygen-reducing site is arranged; how many metal redox-active groups, one or two, are involved. In this regard, the role of haem b_{595} is of special interest. It is the high-spin pentacoordinate haem ligated by His_{19} of subunit I and mediates electron transfer from haem b_{558} to haem *d* (Borisov, Arutyunyan, Osborne, Gennis, & Konstantinov, 1999; Poole & Williams, 1987) with which it forms the binuclear active site. Others have suggested that haem b_{595} serves as a second, additional oxygen-binding site in the enzyme (D'Mello, Hill, & Poole, 1996). A comprehensive understanding of this oxidase will require its crystal structure, which has to date proved elusive.

2.3.3 Cytochrome bd-II

A third oxidase was discovered (Dassa et al., 1991) when a genomic sequence upstream of the *appA* gene (encoding an acid phosphatase) was found to comprise two cotranscribed genes (*appCB*) encoding a putative third oxidase, homologous to cytochrome *bd-I*. These genes (*appC* and *appB*) were predicted to encode proteins with an M_r of 42.3 and 58.1 kDa with hydropathy patterns similar to those of cytochrome *bd-I*. Purification of the proteins (Sturr, Krulwich, & Hicks, 1996) revealed a complex able to catalyse oxygen uptake and with spectral properties indistinguishable from cytochrome *bd-I*, making its contributions to *E. coli* physiology difficult to discern.

The contribution of cytochrome *bd-II*, as it is now called (also AppBC) to cell physiology has been controversial. Based on growth properties of a strain lacking the proton-translocating NADH dehydrogenase I (NDH-I, encoded by the *nuo* operon), cytochrome *bo′* and cytochrome *bd-I*, so that electron transfer is confined to a route from the non-proton-translocating NADH dehydrogenase II (NDH-*II*, encoded by *ndh*) to cytochrome *bd-II*, and a quantitative physiological analysis, Bekker, de Vries, Ter Beek, Hellingwerf, and de Mattos (2009) concluded that cytochrome *bd-II* does not contribute to the generation of a PMF ($H^+/O = 0$). However, Shepherd, Sanguinetti, Cook, and Poole (2010) had proposed that the oxidase might contribute to a PMF by consuming protons during the synthesis of γ-aminobutyric acid. A subsequent study in which proton translocation was examined directly using oxonol VI or acridine orange fluorescence in everted membrane vesicles revealed; however, the formation of a $\Delta\Psi$ (inside positive) and internal acidification of the vesicles (Borisov, Murali, et al., 2011). This proton movement would correspond to proton ejection to the periplasm in intact cells. Furthermore, spheroplasts from the same strain gave an $H^+/O=1$ in oxygen pulse ('Mitchell') experiments. Finally, microsecond resolution single-turnover electrometry after reconstitution of the purified *bd-II* oxidase into liposomes showed that the molecular mechanism of oxidase turnover and PMF generation appeared similar to that of cytochrome *bd-I* (Borisov, Murali, et al., 2011).

The apparent discrepancy between the growth physiology experiments and direct measurements of proton movements has been resolved by Sharma, Hellingwerf, de Mattos, and Bekker (2012). It is due to the activities of two alternative catabolic pathways—the pyruvate oxidase pathway for acetate production and a pathway with methylglyoxal as an intermediate for the production of lactate. Since the ATP yields of these pathways are lower than the energy-coupled routes assumed to be operative during the energy-limited growth conditions used by Bekker, this led to erroneous calculation of the catabolic rate of ATP synthesis and the conclusion that cytochrome *bd-II* did not contribute. A detailed critique of the papers by Bekker and Sharma is given in the latter.

Thus, in conclusion, our current knowledge of the structure and function of cytochromes *bd-I* and *bd-II* suggests that their functions should be interchangeable, yet they appear to be differently regulated, with *cydAB* being maximally expressed at 56% AU and *appCB* being maximally expressed at 0% AU.

2.4. Oxygen relationships

The three terminal oxidases of *E. coli* have distinct relationships with oxygen. Over several decades, researchers have measured the apparent affinities of cytochromes *bd* and *bo′*. All these measurements have shown that cytochrome *bd* has an oxygen affinity significantly higher than cytochrome *bo′*. However, the values for K_m reported diverge widely, due in part to the insensitivity of membrane-covered oxygen electrodes for measuring the very low-oxygen concentrations (below 1 μM) necessary for such high-affinity systems. A number of better methods are available, including the oxygen dependence of luminescence from *Photobacterium* cells (Poole, Lloyd, & Chance, 1979) but these alternatives have not been widely used. The most useful and widely adopted is the use of myoglobin or leghaemoglobin as reporter of dissolved oxygen (Bergersen & Turner,

1979). In essence, the formation of the oxygenated globin (oxymyoglobin or oxyleghaemoglobin, respectively) is dependent on the (known) dissociation constants for oxygen of these globins. The method involves preparation of the oxygenated globin, which is incubated in an anoxic buffer with bacterial cells, membranes or purified oxidase. Upon provision of an oxidizable substrate, respiration proceeds using the low-oxygen concentrations made available by dissociation of the oxygenated globin. The deoxygenation can be recorded in real time and continuously by exploiting the marked spectroscopic differences between the oxygenated form and the deoxygenated (ferrous) form. Precautions are taken to ensure: (i) that the absorbance measurements at the selected wavelengths are unaffected by turbidity of cells or membranes, (ii) that the measured affinities are independent of cell or membrane concentration and (iii) that the observed absorbance changes are due to deoxygenation, not globin oxidation. The progress of deoxygenation is used to calculate the fractional oxygenation of the globin, from which a plot of oxidase velocity versus oxygen concentration is constructed and thence Lineweaver–Burk and Eadie–Hofstee plots. A fuller explanation is given in D'Mello et al. (1996).

Application of this method to *E. coli* has not only confirmed the very high oxygen affinity of cytochrome *bd-I*, as shown in Table 2.1, but also revealed an affinity much higher than the older one, derived from electrode-based methods. Note that the affinity of the *bd*-type oxidase of *Klebsiella pneumoniae* is also very high but that in *Azotobacter vinelandii* the oxidase has a low affinity (high K_m) but (not shown) an exceptionally high turnover rate. These properties may contribute to the role of this oxidase in rapid oxygen consumption necessary to provide respiratory protection for the anoxic process of nitrogen fixation in this aerobe (Poole & Hill, 1997).

Cytochrome *bd* has long been considered to have an exceptionally high affinity, with K_m values as low as 3–8 nM being reported for respiring cells and membranes of *E. coli* in which cytochrome *bd-I* was the dominant oxidase (Table 2.1). However, the near identity of the spectral signatures of cytochromes *bd-I* and *bd-II* does not allow reliable conclusions to be drawn about the oxygen affinity of the latter. Measurements of the K_m of cytochrome *bd-II* for oxygen were published recently (Jesse et al., 2013) but were performed using a membrane-covered Clark electrode with inadequate sensitivity in the nanomolar range. The value cited for cytochrome *bd-II* (2 μM) was almost 10-fold higher than the value reported recently, obtained using the globin deoxygenation technique (Table 2.1). Furthermore, the K_m for cytochrome *bd-I* was almost 100-fold higher than previous

Table 2.1 Oxygen affinities of selected bacterial terminal oxidases as determined by the deoxygenation of globins

Bacterium	Oxidase (cytochrome)	K_m (O_2) (μM)	References
E. coli	*bd-I*	0.003–0.008	D'Mello et al. (1996)
	bd-II	0.24	Jesse et al. (2013)
	bo′	0.016–0.35	D'mello et al. (1995)
Azotobacter vinelandii	*bd*	4.5	D'Mello, Hill, and Poole (1994)

The three terminal oxidases of *E. coli* have distinctive properties. It is tacitly assumed that these differences in affinity for oxygen (above), contributions to proton translocation (above), and patterns of expression as a result of transcriptional regulation (e.g. Tseng, Albrecht, & Gunsalus, 1996) are reflected in their physiological roles and fitness for particular environmental niches. Numerous studies have reached the consensus view that the *cyo* genes encoding cytochrome bo′ are up-regulated under conditions of optimal oxygen availability whereas up-regulation of the *cydAB* genes encoding cytochrome *bd-I* is observed on oxygen limitation or in the presence of competing haem ligands such as NO, CO and cyanide. Indeed, cytochrome *bd-I* is reported to protect the cell from various environmental stresses such as cyanide (Ashcroft & Haddock, 1975), azide and divalent metal ions (Poole, Williams, Downie, & Gibson, 1989) and cytochrome *bd-I* null mutants are more sensitive to hydrogen peroxide than are wild type cultures (Lindqvist, Membrillo-Hernández, Poole, & Cook, 2000). Significantly, cytochrome *bd-I* has been shown to protect *E. coli* from NO-induced growth inhibition (Mason et al., 2009). Explanations have been sought in the ligand (NO, CO)-binding kinetics of cytochromes *bd-I* and *bo′* but these do not fully explain the observed physiological properties (discussed by Jesse et al., 2013).

determinations using the globin deoxygenation method (Table 2.1). Both papers agree, however, on the much higher oxygen affinity of cytochrome *bd-I*. The values published for assays that have used the sensitive globin deoxygenation method are summarized in Table 2.1. The apparent conflict between maximal expression of *appCB* at 0% AU and the relatively low affinity of the encoded oxidase for oxygen does not suggest a function for cytochrome *bd-II* and highlights the deficits in our knowledge of fundamental physiology even for one of the best studied organisms on Earth.

3. EFFECTS OF OXYGEN AVAILABILITY ON *E. COLI* GENE EXPRESSION

3.1. Transcriptional reprogramming and inference of TF activities under stable (steady-state) and dynamic conditions (transitions)

Transcriptome analysis provides an insight into the cellular responses that are elicited by exposure to different oxygen concentrations. As stated above, whilst much is known about *E. coli* gene expression and metabolism under fully aerobic and anaerobic conditions, the intermediate oxygen tensions that are encountered in natural niches are under studied. Thus, the first and simplest transcriptomic approach used in the SUMO project was to determine the effects of oxygen availability on the transcript profiles of steady-state cultures maintained at fixed points across the aerobiosis scale (Rolfe et al., 2011). However, bacteria rarely enjoy the security provided by stable sustainable environments; enteric bacteria, like *E. coli*, are frequently exposed to rapid changes in oxygen availability as they enter and leave a host. Therefore, a second approach, in which cultures were monitored during transitions from oxygen-replete to oxygen-starved conditions and vice versa, was employed to mimic these 'relays' of naturally occurring environmental shifts (Rolfe et al., 2012). Application of these two approaches to highly controlled chemostat cultures, where it is possible to alter a single environmental parameter, in this case oxygen availability, enabled the capture of complementary transcriptomic datasets that could be integrated with other physiological measurements made within the consortium.

3.1.1 TFInfer: Inferring TF activities from transcriptomic datasets

A systems level understanding of gene regulation ultimately requires knowledge of TF activities; unfortunately, simultaneous direct measurement of the intracellular activities of numerous TFs is not technically feasible. However, probabilistic modelling of transcript profiles can be used to reveal 'hidden' information present in such data by simultaneously inferring the activities of the TFs that control the activity of RNA polymerase. In order to extract information on TF activity from transcriptomic data, a statistical model was developed. The model (Sanguinetti, Lawrence, & Rattray, 2006) assumes a log-linear approximation for the transcriptomic response, whereby changes in (log) expression of genes are modelled as a linear combination of the change in activity of their regulators. Whilst this undoubtedly represents an approximation to a complex dynamical process, it captures the first-order behaviour of the dynamical system. Furthermore, this assumption is essential in ensuring computational feasibility, thereby enabling prediction of TF activities on a large scale. TF activity itself is not observed: the framework treats it as a latent variable whose distribution can be evaluated *a posteriori* via the application of Bayes' theorem. In formulae, the basic model is summarized as

$$y_n(t) = \sum_m X_{nm} b_{nm} c_m(t) + \varepsilon_n \tag{2.1}$$

where n is an index over genes, m is an index over TFs, t indexes different conditions (or time points for temporal data). X is a binary matrix whose nm entry is 1 if and only if TF m regulates gene n. This matrix is assumed to be known and is compiled using curated data from repositories such as EcoCyc (Keseler et al., 2013) and RegulonDB (Salgado et al., 2013), which provide binding information for 167 TFs targeting over a thousand genes. The terms b and c are latent (unobserved) random variables representing the response strength of genes to TFs and the activity of TFs, respectively, whilst ε_n is a zero mean, normally distributed error term. Under the Bayesian framework, these latent variables are assigned prior distributions which, together with the model Eq. (2.1), define the joint distribution of the observations and the model variables. The response strength variables (b) are assigned a zero mean Gaussian to encode equal prior probability of activation and repression. The choice of the prior distribution over the TF activities depends on the experimental design: in a dataset consisting of multiple steady-state measurements, an independent Gaussian prior for the TF activity at each oxygen level seems a reasonable assumption. On the contrary, when modelling time series data, it is important to capture the dynamical nature of the data; temporal continuity is therefore ensured through the choice of a Markovian random walk prior.

Posterior estimation in model (2.1) is intractable (when jointly considering both response strengths and activities); Sanguinetti et al. (2006) provided an approximate inference algorithm which enables efficient inference by enforcing a posterior independence *ansatz*. The model has been recently released as an open source C# package, TFInfer (Asif et al., 2010), which provides a user friendly graphical front end to the computational modules and enables use by non-experts.

3.1.2 Gene expression and TF activities under steady-state conditions

Reprogramming *E. coli* metabolism in response to oxygen availability has been associated with widespread changes in gene expression. Several gene expression studies that compare aerobic and anaerobic cultures of *E. coli* grown on various different media have been reported (Constantinidou et al., 2006; Kang, Weber, Qiu, Kiley, & Blattner, 2005; Salmon et al., 2003). Based on these studies, it has been suggested that up to ~500 and perhaps as many as ~1450 of the genes that are expressed under aerobic conditions have altered expression under anaerobic conditions. However, when steady-state cultures grown with excess oxygen (217% aerobiosis, as defined above) were compared with anaerobic (0% aerobiosis) steady-state cultures under SUMO conditions only 63 genes were significantly regulated (Rolfe et al., 2011). These data can be interpreted as representing the minimal set of differentially expressed genes that distinguish aerobic and anaerobic growth, without the many artefacts that are caused by differences in growth rate and stress that are inherent in batch growth (Hoskisson & Hobbs, 2005). This clearly illustrates the benefits of using chemostat culture to grow the bacteria under defined and controlled physiological conditions for transcriptomic analyses. The critical adaptive responses appear to be enhanced expression of TCA cycle and glyoxylate shunt genes under aerobic conditions, with the cyclical form of the TCA cycle being restored as oxygen availability increased by expression of the operon encoding succinate dehydrogenase (SDH)-succinyl CoA synthetase. Induction of the glyoxylate shunt under aerobic conditions suggests that initially the bacteria might not fully oxidize the substrate glucose, but continue to excrete a proportion of the carbon supplied as acetate, which is however undetectable in the culture medium because it is 'immediately' taken up via the acetate transporter ActP (induced >10-fold under aerobic conditions) for metabolism via the glyoxylate cycle. The switch to aerobic respiratory growth was completed by the synthesis of the relatively low-oxygen affinity terminal oxidase cytochrome *bo′* and down-regulation of transcripts associated with anaerobic fermentation (Fig. 2.1). These transcriptional changes are consistent with the relief of ArcA-mediated repression and inactivation of FNR (see below).

Comparison of the complete set of transcript profiles obtained at pre-determined points in the microaerobic–aerobic range with that of the anaerobic steady-state indicated that a total of 89 genes exhibited expression changes (Rolfe et al., 2011). Transcripts with altered abundance were associated with central metabolic functions (39%), unknown function (11%), transcription regulation (7%), outer membrane proteins and adhesions (5%), and the oxidative stress response (4%) (Rolfe et al., 2011). Stepwise comparison of the transcriptome changes at each steady state suggested a progressive switch to aerobic respiratory metabolism (Fig. 2.1). The *cydAB* operon (encoding cytochrome *bd-I*, Section 2.3.2) was maximally expressed at 56% AU (6.4-fold enhanced compared to 0% AU), whereas the *cyoABCDE* operon (encoding cytochrome *bo′*, Section 2.3.1) was maximally expressed under fully aerobic conditions (7.2-fold enhanced compared to 0% AU), and *appCB* (encoding cytochrome *bd*-II) was maximally expressed at 0% AU (1.4-fold compared to 115% AU) (Fig. 2.1). Cytochrome *bd-I* has an extremely high-oxygen affinity (K_m 3–8 nM) (D'Mello et al., 1996), whilst cytochrome

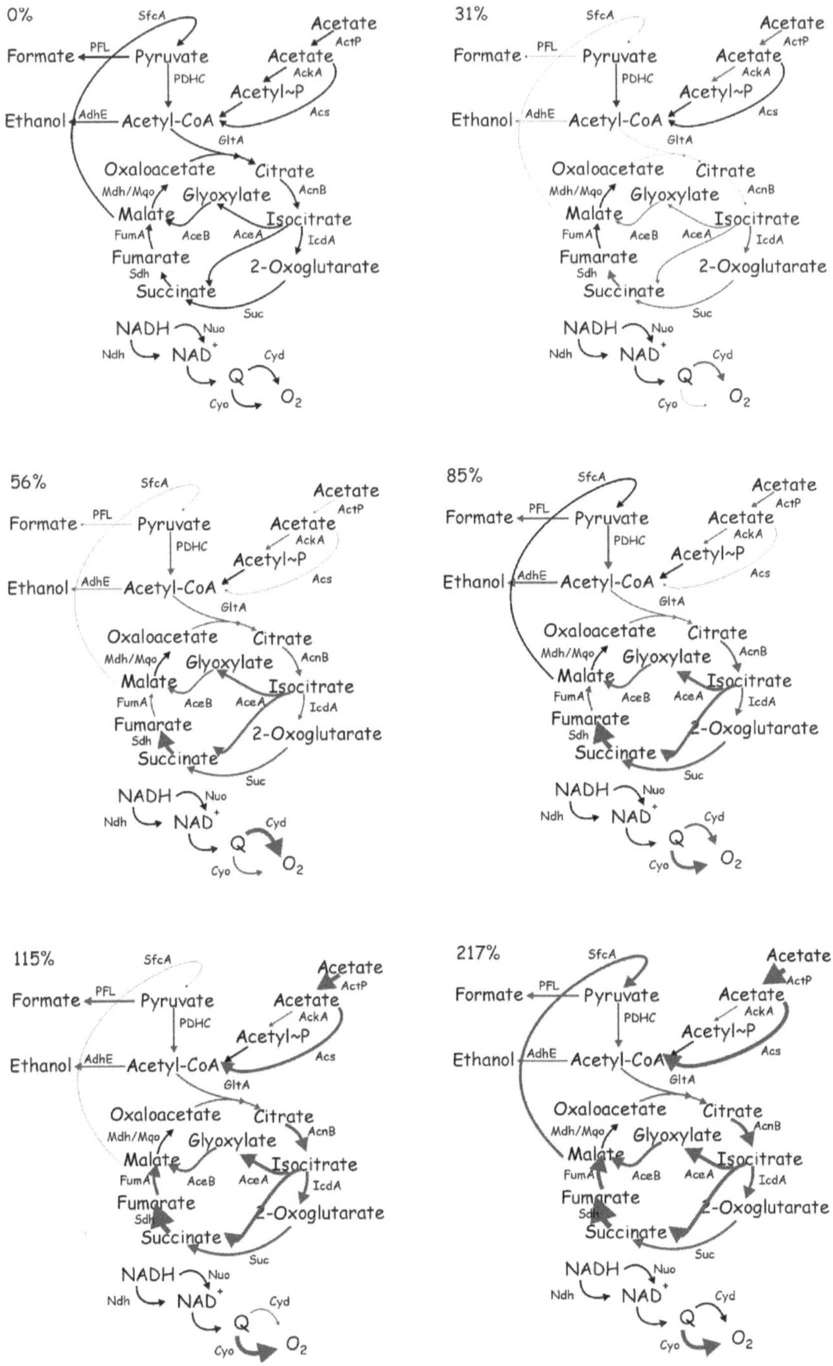

Figure 2.1 Changes in abundance of central metabolic transcripts in steady-state cultures grown at fixed pre-determined points on the aerobiosis scale. Transcripts with decreased abundance relative to the anaerobic steady state (0% aerobiosis) are indicated by dashed arrows; transcripts with increased abundance relative to the anaerobic steady state (0% aerobiosis) are indicated by solid arrows. The width of each arrow provides an indication of the change in transcript abundance—the wider the arrow that greater the change. The point on the aerobiosis scale for each panel is indicated. (See the color plate.)

bo' has a relatively low affinity for oxygen (K_m 16–350 nM) (D'mello, Hill, &, Poole, 1995) (Table 2.1). Thus, the expression profiles match the biochemical properties of these oxidases. Furthermore, spectroscopic measurements of the cytochrome *bd* and cytochrome *bo'* content of intact cells from steady-state cultures at different aerobiosis values showed that the changes in abundance of the *cydAB* and *cyoA-E* transcripts correlated well with the changes in the amounts of cytochrome *bd* and cytochrome *bo'* in the cell; good correlation between transcript and protein abundances is a notable feature of the experiments carried out under SUMO conditions (Rolfe et al., 2011; Trotter et al., 2011). Thus, the shifts in oxidase expression are consistent with a progressive switch from fermentation to microaerobic respiration (with cytochrome *bd-I* acting as the terminal oxidase), and then to aerobic respiration (with cytochrome *bo'* as the terminal oxidase), as oxygen availability was increased (Fig. 2.1). However, these experiments did not allow function to be assigned to cytochrome *bd-II*. And also, it can be conclude that post-transcriptional regulation does not play a major role in the expression of the proteins involved in these transitions.

The observed transcript and protein profiles result from changes in the activities of regulators that respond to oxygen itself, or to metabolic and environmental signals that are sensitive to oxygen availability. Although the roles of FNR and ArcA in mediating the changes in transcript profiles necessary for adaptation to growth at different oxygen availabilities are well established (see above), the consequent changes in physiology are expected to modulate the activities of other TFs that respond to specific metabolic signals that are affected by oxygen availability. Probabilistic modelling (TFInfer, see above; Asif et al., 2010) was used to simultaneously infer the activities of 134 TFs, 23 of which responded in at least one of the steady states examined (Table 2.2). All seven of the so-called global regulators (ArcA, CRP, Fis, FNR, H-NS, IHF and Lrp; Martínez-Antonio et al., 2003) of *E. coli* K-12 were amongst these 23 TFs and the signals that modulate the action of these global TFs are all related to the energetic state of the bacterium, and so their responsiveness to changes in oxygen supply is perhaps not surprising. However, more surprisingly the activity of the oxygen-sensing TF, FNR was predicted to decrease rapidly only at aerobiosis values greater than 56% and this was confirmed experimentally, validating the predicted FNR activities from the TFInfer model (Rolfe et al., 2012). In contrast, the activity of the indirect oxygen-sensor ArcBA was predicted to decrease linearly with increasing aerobiosis. Because ArcA phosphorylation is inhibited by oxidized quinone (Georgellis et al., 2001), it was

Table 2.2 Identification of *E. coli* K-12 regulators that respond to oxygen availability

Transcription factor	Signal/function	Condition
Acid resistance/metabolism		
EvgA	Acid resistance	SS
FhlA	Formate	SS, NO, ON
GadW	Acid resistance	NO
GadX	Acid resistance	SS
PdhR	Pyruvate	SS, NO, ON
Amino acid metabolism		
ArgR	L-arginine	NO, ON
CysB	*O*-acetyl-L-serine	NO, ON
LeuO		SS
Lrp	L-leucine	SS, NO, ON
MetJ	*S*-adenosylmethionine	NO, ON
TdcA	Threonine/serine metabolism	ON
TrpR	Tryptophan	ON
Carbon metabolism		
CRP	Cyclic AMP	SS, NO, ON
DcuR	C_4 dicarboxylates	ON
DgsA	Functional state of PTS system (binds EII)	NO, ON
FadR	Acyl-CoA	ON
GlcC	Glycolate	SS, ON
HcaR	Hydrocinnamic acid	ON
MalT	Maltotriose	NO, ON
RbsR	D-ribose	SS
SgrR	Glucose-phosphate stress	ON
Regulation of anaerobic metabolism		
ArcA	Redox state of quinone pool, fermentation products	SS, NO, ON
FNR	Oxygen	SS, NO, ON

Table 2.2 Identification of *E. coli* K-12 regulators that respond to oxygen availability—cont'd

Transcription factor	Signal/function	Condition
Metal ion management		
CueR	Cu(I)	ON
CusR	Anaerobic Cu(II)	SS, NO, ON
Fur	Iron starvation	SS, NO, ON
IscR	Iron–sulphur cluster depletion	SS, NO, ON
ModE	Molybdate	SS, NO, ON
NikR	Nickel	NO, ON
ZraR	Zinc	ON
Cell envelope		
CpxR	Extracytoplasmic stress, mis-folded proteins	SS, NO, ON
FlhDC	Unknown	SS, ON
PspF	Extracytoplasmic stress	SS
Oxidative stress		
OxyR	Peroxide stress	NO
SoxR	Redox stress (superoxide)	NO, ON
SoxS	Redox stress (superoxide)	NO
Nucleoid proteins		
Fis	Growth phase	SS, NO, ON
H-NS		SS, NO, ON
IHF		SS, NO, ON
PhoB	Inorganic phosphate	ON
Others		
BirA	Biotinoyl-5′ adenylate	NO
NarL	Nitrate	SS, NO, ON
NrdR		SS, NO, ON
PurR	Hypoxanthine	ON

The activities of 166 regulators were inferred from the transcript profiles of steady-state cultures grown at 0%, 31%, 56%, 85% and 115% aerobiosis. Responsive transcription factors were defined as those exhibiting a signal-to-noise ratio >5. In the column headed *Response*, + indicates increased activity as aerobiosis increases; − indicates decreased activity as aerobiosis increases; x indicates a more complex activity profile with maxima or minima at intermediate aerobiosis values. The conditions under which each TF responds are indicated; SS (steady-state cultures at different points on the aerobiosis scale), NO (anaerobic to aerobic transition), ON (aerobic to anaerobic transition).

expected that the quinone pool would be become increasingly oxidized as the level of aerobiosis increased, but this was not the case (Rolfe et al., 2011). Subsequent measurement of the amount of phosphorylated ArcA revealed a good correlation with the predicted ArcA activities and with aerobiosis (Rolfe et al., 2011). These data indicated that the relationship between the redox state of the quinone pool and ArcBA activity is not as simple as was initially envisaged. One source of complexity is that *E. coli* uses three different types of quinone: ubiquinone, dimethyl-menaquinone and menaquinone (of which the last two have a precursor/product relation). Significantly, these three quinones display an increasing negative redox mid-point potential, which is (partially) reflected in the specificity of the reactions in which they are involved (possibly to minimize the redox 'mismatch' with the enzymes with which they react) (Schoepp-Cothenet et al., 2009; Unden & Bongaerts, 1997). Construction and analysis of a set of *E. coli* mutants defective in their quinone biosynthesis pathways (Alvarez, Rodriguez, & Georgellis, 2013; Bekker et al., 2010; Sharma, Stagge, Bekker, Bettenbrock, & Hellingwerf, 2013) revealed that a mutant with ubiquinone only, with dimethyl-menaquinone only, and a mutant containing a mixture of the two menaquinones only (i.e. deficient in ubiquinone synthesis) all showed clear active/inactive ArcA transitions upon modulation of the rate of oxygen supply to the cells (but note the contrasting results of Alvarez et al., 2013) and which might reasoned in unknown differences between strains. Whether or not menaquinone itself can achieve this modulation is not known because mutants in *E. coli* that exclusively contain this latter quinone are not yet available.

The possibility that both ubiquinone and menaquinone modulate ArcB activity was first suggested by Bekker et al. (2010), who observed that ArcB activity in the non-SUMO strain *E. coli* BW25113 did not decrease linearly as the oxygen supply increased, as was found for the SUMO strain *E. coli* MG1655 (see above). Instead, upon allowing a limited rate of oxygen supply to the culture the ArcB activity rapidly decreased, until at ~40% AU its activity increased again, to reach a maximum at around 80% AU, after which it steeply decreased to zero at 100% AU (see Alexeeva et al., 2003). This experiment has been interpreted in the following way (Fig. 2.2): anaerobically grown *E. coli* BW25113 cells contain predominantly menaquinone, which, because of its strongly negative mid-point potential, leads to rapid inactivation of ArcB. When the % AU increases, the menaquinone content decreases and the amount of ubiquinone strongly increases (Bekker et al., 2010). This presumably leads to a situation (at around 80% AU) in which

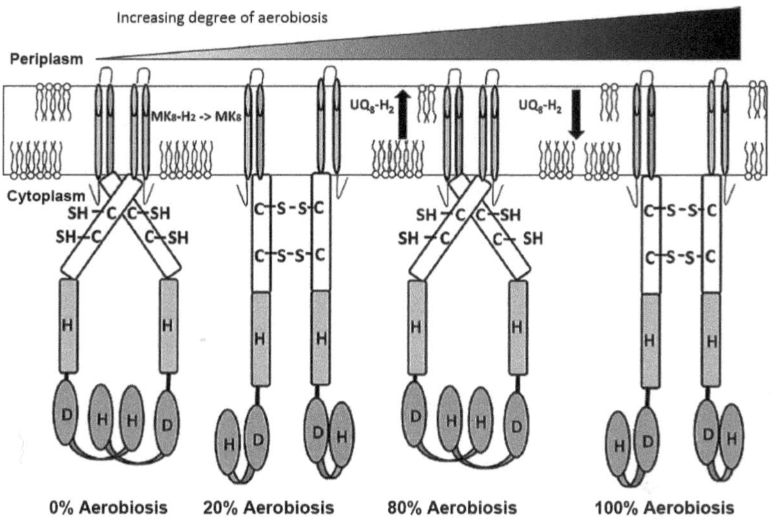

Figure 2.2 Simplified model for modulation of ArcB activity, in a transition from fully anaerobic conditions to conditions with oxygen saturation, by the combined action of ubiquinone and the menaquinones present in the cytoplasmic membrane of *Escherichia coli*. For further explanation: see text and Bekker et al. (2010).

ubiquinol (i.e. reduced ubiquinone) and (oxidized) menaquinone co-exist in the cytoplasmic membrane. Their concentrations then will be such that ubiquinol overrides the inhibitory effect of menaquinone (Sharma, 2013). Alvarez et al. (2013) also report the mid-point potential of (the two?) cysteines of ArcB: −41 mV and seem to suggest that this would make it impossible for ubiquinol to reduce ArcB. This, however, need not be true if the actual redox potential of ubiquinone/nol can drop below zero (i.e. at very high-ubiquinol/ubiquinone ratios). In agreement with this (Sharma et al., 2013) report comparable on/off transitions of ArcB activity in *E. coli*, strains that contain either exclusively ubiquinone or exclusively demethylmenaquinone. A stringent test of this model, beyond demonstrating that both ubiquinone/nol and menaquinone/nol can modulate ArcB activity, is not possible (yet) because a reliable assay of the redox state of menaquinone is not available. It is also relevant to note that the relation between ArcB activity and the % aerobiosis can be much simpler in other *E. coli* strains (Rolfe et al., 2011).

Taken together, it is clear that there are still questions to be answered concerning the mechanisms of ArcBA activation and inhibition with roles for ubiquinone, menaquinone and allosteric effects of fermentation products (e.g. acetate) all contributing to determining ArcA activity (Bekker et al., 2010; Georgellis, 1999; Rodriguez et al., 2004; Sharma et al., 2013).

3.1.3 Metabolic fluxes change with aerobiosis

The changes in TF activities and the consequent changes in gene expression that occur in response to oxygen availability result in the reprogramming of central metabolism. The point on the AU scale at which an *E. coli* culture is maintained not only affects acetate production (the basis of the aerobiosis scale, Section 2.1) but also the concentrations of other fermentation products. Ethanol, formate and lactate production decreased with increasing aerobiosis, but production of these fermentation products stops at aerobiosis levels around 50% AU (Steinsiek, Frixel, Stagge, & Bettenbrock, 2011). This behaviour shows that, at intermediate aerobiosis levels, fermentation and respiration co-exist and that with increasing availability of oxygen the cells rely more and more on respiration for regeneration of NAD^+, consistent with the transcriptional reprogramming discussed above (Section 3.1.2). This trend was also reflected in biomass yield, which increased gradually with increasing aerobiosis. At low-oxygen concentration also in glucose-limited chemostat cultures overflow metabolism, that is, the production of acetate at aerobic conditions can be observed with acetate being the only by-product produced. As acetate production does not contribute to NAD^+ regeneration, its production reflects the fact that only part of the acetyl-CoA produced can be fed into the TCA cycle either because some enzymes of the TCA cycle are not available in sufficient amounts or because the respiratory chain cannot operate at sufficient speed to oxidize the additional NADH produced by the TCA cycle. This trend is also reflected in the intracellular concentrations of some metabolites, reflecting the observation that, with increasing aerobiosis, the fluxes through the TCA cycle increase gradually (S. Stagge and K. Bettenbrock, personal communication).

By switching from aerobic to anaerobic metabolism or more precisely by switching from respiration to fermentation, extensive changes in the central metabolism of *E. coli* take place. These are most pronounced in the TCA cycle that shifts from a cyclic into a branched mode. In the branched mode, the TCA still serves to produce biosynthesis precursors but no longer produces NADH for respiration. The switch to the branched mode is reflected by a very strong regulation of SDH that is maximally repressed at anaerobic conditions due to the action of the global regulators ArcBA and FNR (Park, Tseng, & Gunsalus, 1995; Shen & Gunsalus, 1997). As regulation especially by the ArcBA system and metabolism are tightly connected, mutations in enzymes of the central metabolism are supposed to change the behaviour of *E. coli* at different oxygen concentrations. In *E. coli*, SDH catalyses the conversion of succinate to fumarate coupled to the reduction of ubiquinone. The reverse reaction that is active in the production of succinate as part of mixed-acid fermentation is catalysed by the fumarate reductase (FRD) complex. In this case menaquinone is used as electron donor. SDH and FRD show high similarities and are able to complement mutations in the other enzyme if their expression is modified (Cecchini, Schröder, Gunsalus, & Maklashina, 2002; Maklashina, Berthold, & Cecchini, 1998). Unlike expected deletion strains lacking SDH grew similar to the wild type under aerobic, microaerobic and anaerobic conditions even without such modifications (Steinsiek et al., 2011). SDH^- and an FRD^- grew well at all aerobiosis levels. Whilst the FRD^- strain displayed an aerobiosis scale indistinguishable from the wild type the SDH^- strain showed a different behaviour. Acetate production in this strain was lower at aerobiosis levels from 0% to 80% but from 80% on no further reduction in acetate production occurred. This strain hence displays a drastically increased overflow metabolism, obviously due to reduced fluxes from succinate to fumarate.

3.2. Mathematical models describing steady-state gene regulation and metabolic fluxes

As described in the above sections, the oxygen response comprises changes in TF activities, gene expression and metabolic fluxes. The observed response results from a mutual interplay of these levels. Oxygen availability has major effects on energy and redox metabolism, either directly via oxygen-dependent or oxygen-inhibited metabolic reactions or via the control of TFs. In this way, oxygen availability determines the uptake and distribution of the carbon substrate glucose and biomass growth. Changes in carbon metabolism influence redox balance and lead to changes in transcriptional regulation. In a chemostat, the supply of carbon substrate and electron acceptor is constant, and the amount of biomass determines the specific uptake fluxes. In this way, the amount of biomass influences the molecular processes of metabolism and regulation. A systems level study of these complex interdependencies requires comprehensive measurement data at all relevant levels, as discussed in the preceding sections of this review. However, what is still missing is an integrated framework that allows the analysis of the experimental data in the context of a multi-level network model. We present a modelling framework capturing the interplay between the different levels that participate in producing the overall oxygen response.

The model consists of four interlinked parts: extracellular substances, metabolism, genetic regulation and growth.

Extracellular substances follow the chemostat equations

$$\frac{dc_i}{dt} = c_X J_i - c_i D + J_{ex,i}$$

where c_i is the concentration of the extracellular substance and c_X is the biomass concentration (Smith & Waltman, 2008). The parameters of this part of the model are the dilution rate D and the in-/outflow from the chemostat $J_{ex,i}$. For the gaseous compounds (oxygen and carbon dioxide), the exchange flux is $J_{ex,i} = k_{in,i} - k_{ex,i} \times c_i$, where $k_{in,i}$ is the supply rate and $k_{ex,i}$ is a coefficient that describes the outgassing. For the other compounds, material exchange occurs solely via the liquid phase and $J_{ex,i} = Dc_{in,i}$. The flux J_i is the corresponding production (positive) or uptake (negative) flux, which is determined by the metabolism model.

Central metabolism is modelled using the thermodynamic-kinetic modelling (TKM) formalism. The use of TKM assures that the model equations follow thermodynamic constraints even in situations where metabolic cycles exist (Ederer, 2010; Ederer & Gilles, 2007). In TKM, every compound with concentration c_i has a thermokinetic potential ξ_i which are related via the capacity C_i, such that $c_i = C_i \times \xi_i$. The capacities are calculated from the chemical standard potentials (Ederer, 2010; Ederer & Gilles, 2007) that are tabulated for most relevant metabolites of central metabolism (Alberty, 2003, 2006). The thermokinetic potentials determine the thermokinetic driving force F_j for a reaction j. For example, the reaction $A + B \rightleftharpoons C$ has the thermokinetic force $F = \xi_A \times \xi_B - \xi_C$. Finally, the reaction rate is determined by the law $R_j(\xi) J_j = F_j$, where in general $R_j(\xi)$ is a positive function of the thermokinetic potentials. The choice of $R_j(\xi)$ determines the kinetic law of the reaction j. Realistic, *in vivo* kinetics, are difficult to obtain and *in vitro* data are not always reliable (Kettner & Hicks, 2005). It is assumed that when considering thermodynamic constraints on reaction networks that the coherent consideration of the constraints is more important than the precise reproduction of the kinetic laws and that one velocity parameter (here: the thermokinetic resistance) per reaction is sufficient for describing network behaviour, see for example, Beard (2005) and Klamt, Grammel, Straube, Ghosh, and Gilles (2008). In the present model, a constant R_j is used, except for selected reactions, where the explicit consideration of saturation and activation/inhibition is necessary. The resistances are adjusted such that realistic flux distributions are achieved.

Genetic regulation is modelled by phenomenological functions describing TF activities and gene expression rates. The activity of TFs is assumed to be between 0 (minimal activity) and 1 (maximal activity) and is described by a Hill-type laws, dependent on the respective metabolic signals. For example, FNR activity is assumed to be a function of oxygen concentration. By choosing these activity functions, different hypotheses concerning the metabolic signals of for example ArcBA can be tested and checked for their consistency with experimental data. Expression of metabolic genes is modelled by the differential equation:

$$\frac{dc_{E,i}}{dt} = J_{syn,i} - c_{E,i}\mu$$

where $c_{E,i}$ is the concentration of the enzyme i in arbitrary units, μ is the specific growth rate and $J_{syn,i}$ is the synthesis rate that depends on the TF activities.

The *specific growth rate* of the population is modelled as function of the concentrations of the metabolites that are precursor for biosynthesis in a linear-log fashion. For a description of lin-log kinetics, see for example Visser and Heijnen (2003).

Figure 2.3 shows the results of a steady-state model simulation of *E. coli* K-12 MG1655 in comparison to measurement values for different values of aerobiosis. Qualitative and quantitative features of steady-state flux distributions can be reproduced by the model. This stresses the fact that the flux distributions are largely determined by thermodynamic constraints and justifies the phenomenological but thermodynamically consistent approach for modelling the metabolism. In some cases, metabolite curves are not reproduced well, suggesting either missing regulatory interactions or measurement error. Gene expression of many genes can be reproduced realistically suggesting the fact that the major genetic regulatory systems relevant for the oxygen response of the wild type are known. TF activities were only measured for ArcA and FNR and are in agreement with the model output.

A true quantitative systems level understanding of the oxygen response of *E. coli* is not yet reached and will require the correct description of not only the wild type but also of mutant strains. The present model provides a first step towards a thermodynamically consistent view of the interactions between metabolites, enzymes and TFs of *E. coli* in a steady-state chemostat.

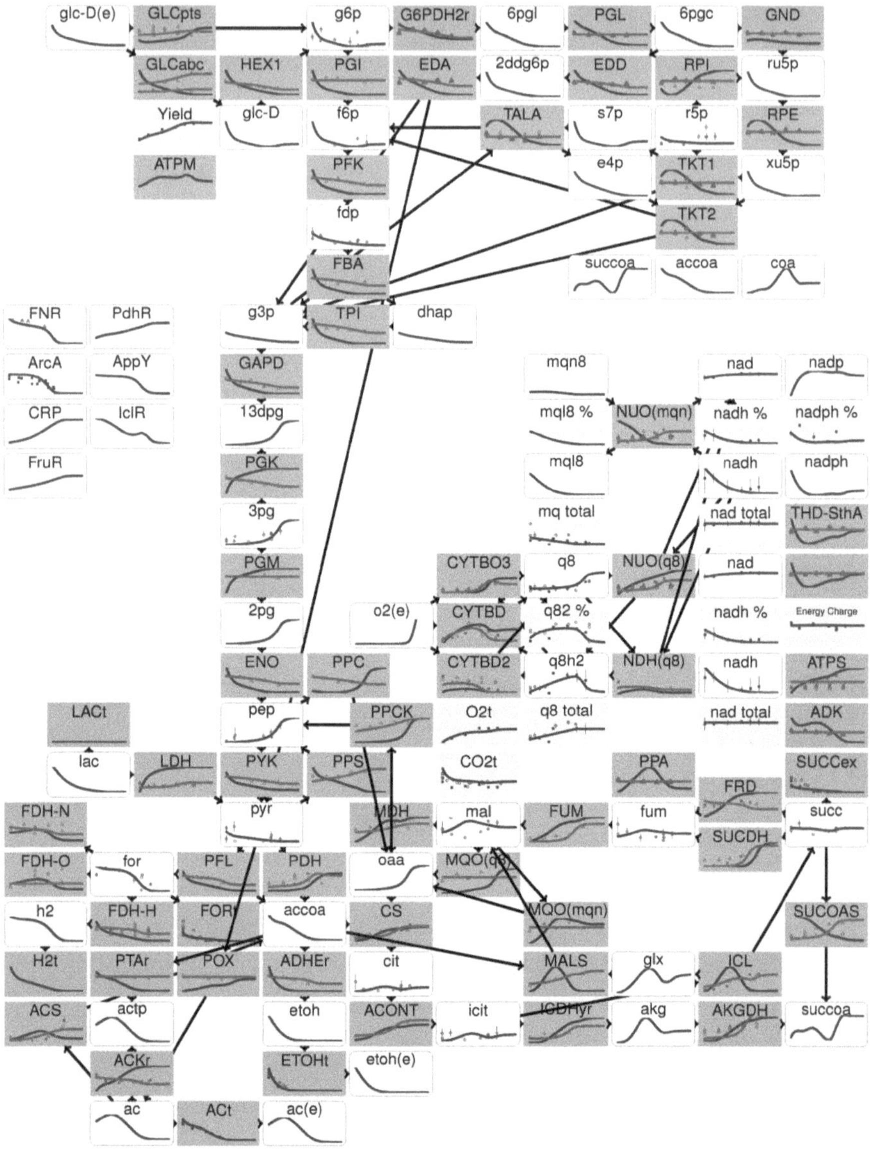

Figure 2.3 Simulation results of the central metabolism model. Shown are glucose uptake and glycolysis (upper left part of the figure), the fermentation pathways (lower left part), the citric acid cycle (lower right part), the electron transport chain (middle right part) and the pentose phosphate pathway (upper right part). White boxes represent metabolites and transcription factors. Dark grey lines and symbols in white boxes show simulations and measurements of concentrations, respectively. The labels of metabolites follow (Reed, Vo, Schilling, & Palsson, 2003): Grey boxes represent reactions. Dark grey lines and symbols in grey boxes show metabolic fluxes. Light grey lines and symbols in grey boxes show gene expression levels. The labels of the reactions follow the lables of (Reed et al., 2003). The variables mql8%, q8h2%, nadh% and nadph% represent the degree of reduction of the menaquinone, ubiquinone, NADH and NADPH pools, respectively. The variables mq total, q8 total, nad total, nadp total are the total concentrations (reduced + oxidized) of the respective redox pairs. The variables O2t and CO2t are the biomass-specific exchange rates of oxygen and carbon dioxide, respectively. (See the color plate.)

3.3. Gene expression and TF activities during transitions

Transitions between oxygen-replete and oxygen-deficient environments (Partridge, Scott, Tang, Poole, & Green, 2006; Rolfe et al., 2012; Trotter et al., 2011) mimic real-life environmental conditions allowing adaptation to new oxygen concentrations to be experimentally investigated. Transcriptomic analysis of the acute phase (first 20 min) of transitions showed that 396 genes (8.1% of the genome) exhibited expression in an anaerobic to aerobic switch and 590 (12.1% of the genome) in an aerobic to anaerobic switch (Rolfe et al., 2012). Thus, the abundances of many more transcripts change during transitions than were observed in simple comparison of aerobic and anaerobic steady-state cultures (see above). This presumably reflects the dynamic nature of the adaptive processes and the many stresses encountered by *E. coli*, whilst oxygen availability is changing. Once metabolism has been reprogrammed, these acute stresses no longer exert their influence on the transcriptome and are thus not apparent from analysis of steady-state cultures.

Of the 396 and 590 genes that exhibited altered expression during the acute phases of the anaerobic to aerobic and aerobic to anaerobic transitions, respectively, 195 and 284 could be classified into 23 functional groups (Fig. 2.4). As expected the major changes in expression were observed for transcripts encoding proteins with roles in central metabolic pathways, including the TCA cycle, oxidative phosphorylation and fermentation (Fig. 2.1). Altered expression of genes associated with the oxidative stress response is presumably indicative of the endogenous oxidative stress experienced in the presence of oxygen. The responses of systems involved in copper ion homeostasis, suggests that the introduction of oxygen alters the redox state of this metal, such that a transcriptional response is necessary. Although a large proportion of the transcriptional reprogramming that occurred during the aerobic to anaerobic transition was essentially the reverse of that of the anaerobic to aerobic transition, transcript changes in the aminosugars, ascorbate, C_5-branched dibasic acids and galactose metabolism functional classes, as well as the translation category, were differentially affected (Fig. 2.4). The transcripts of 204 genes of unknown function (*y* genes) exhibited altered patterns of abundance in the transitions (Fig. 2.5). Of these '*y*' genes, 51 responded in the anaerobic to aerobic transition, 119 in the aerobic to anaerobic transition and 34 in both. These observations highlight the deficits in our knowledge of the mechanisms and processes involved in *E. coli* adaptation to changes in oxygen availability

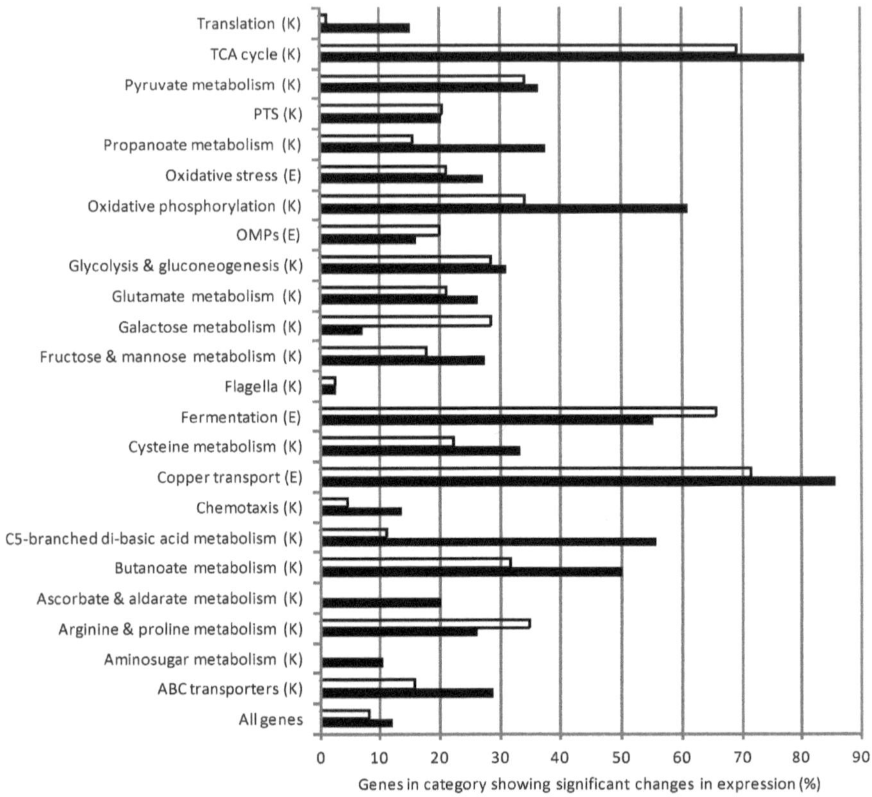

Figure 2.4 Chart showing the proportion of genes in selected functional categories that exhibit altered expression in the first 20 min of the transition to either anaerobic (black bars) or aerobic (white bars) growth. The functional classifications were obtained from KEGG (K) and EcoCyc (E). The values include transcripts that significantly increase or decrease in abundance in response to perturbation.

and emphasize the new insight that can be obtained by perturbation of systems by pointing the way to assigning function to at least some of these genes.

The inference of TF activities in these dynamic environments provides an ideal way of unpicking some of the complex regulatory events that occur during adaptation to changes in oxygen availability. A comparison of the responses of TF activities when aerobic or anaerobic steady states were perturbed revealed the presence of a core set of 16 TFs (including FNR and ArcA) that exhibited activity changes at different fixed oxygen availabilities and in response to introduction/removal of oxygen. This core network contains TFs that respond to oxygen availability and regulate central

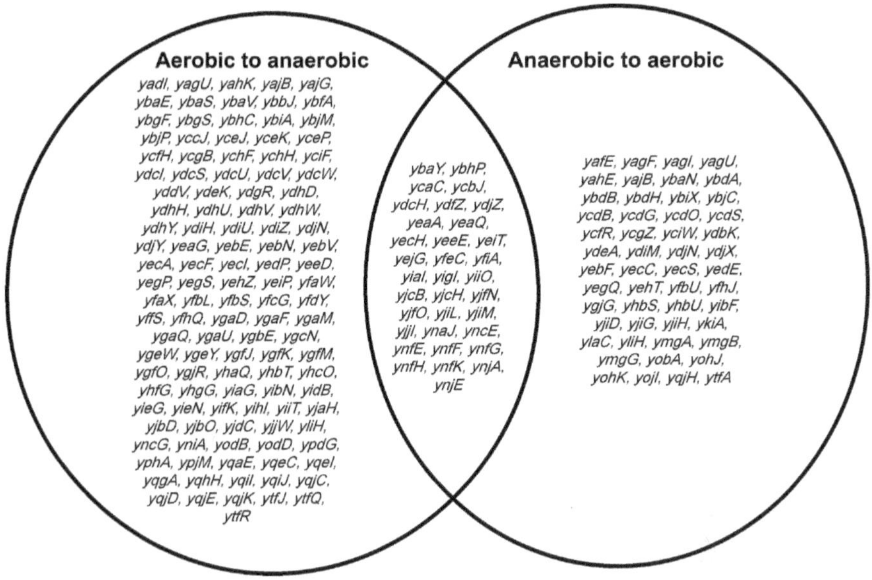

Figure 2.5 Catalogue of genes of unknown function that exhibit altered expression profiles during transitions between environments with different oxygen availability.

metabolism, metal-ion homeostasis, oxidative stress responses, the composition of the nucleoid and amino acid biosynthesis (Table 2.2). Consistent with the observation that the transcriptional reprogramming occurring in the aerobic to anaerobic transition was not simply the reverse of that for the anaerobic to aerobic transition (see above), other TFs exhibited activity changes only during a specific transition (11 TFs for aerobic to anaerobic and 4 for anaerobic to aerobic) or at certain points on the AU scale (5 TFs). Furthermore, of the 44 TFs predicted to respond under one or more of the conditions investigated, 23 were specifically associated with the transitions, whereas only five were specifically linked to the different steady states (Table 2.2). This is perhaps indicative of the additional stresses encountered during the environmental changes occurring during transitions compared to the consistent environmental conditions of the steady-state cultures.

4. SPATIAL ORGANIZATION OF THE *E. COLI* CELL INFLUENCES ArcA AND FNR ACTIVITIES

An interesting feature implied by the transcriptional and protein measurement data described above is the potential influence of spatial effects on the oxygen-responsive TF FNR. The activity predictions based on the transcriptomic data suggested that FNR remains active under microaerobic conditions, even though there is sufficient oxygen to alter cell physiology (Rolfe et al., 2012). The simplest interpretation of this prediction is that the bacterial cytoplasm remains essentially anaerobic under these conditions, presumably due to efficient consumption of oxygen by high-affinity cytochrome *bd-I* (Cyd). FNR activity measured using a *lacZ* reporter assay confirmed FNR activity under anaerobic and microaerobic conditions. This implied that, even on the scale of the bacterial cell, spatial constraints have significant physiological implications, in this case by allowing 'hybrid' metabolism to proceed under microaerobic conditions during which, anaerobic processes can be supported in the cytoplasm, whilst aerobic respiration proceeds in the vicinity of the cell membrane. This concept is consistent with the production of fermentation products (e.g. acetate), induction of the glyoxylate shunt (implying that at least some excreted acetate is utilized) and the absence of detectable oxygen under microaerobic conditions. Furthermore, this could partially account for the need for two TF systems consisting of a membrane-associated sensor interacting with the respiratory chain (ArcBA) and a sensor (FNR) responding to cytoplasmic oxygen, to coordinate global gene expression in response to oxygen availability. To test the hypothesis that spatial constraints are important for oxygen sensing in *E. coli*, a systems approach is essential. Precise experimental measurements of the spatial distribution of oxygen at low concentrations and spatial scales are currently not feasible. Mathematical modelling is therefore a way to test the hypothesis that the cellular locations of the key players play a role in system behaviour. Two complementary modelling approaches were used. In the first approach, the spatial and temporal behaviour of the oxygen diffusion and uptake was analysed by a partial-differential equation system that is based on Fick's laws of diffusion. The major unknown in this approach is the coefficient of oxygen diffusion in the cell and across the cell membrane and the actual oxygen concentration. The model allows deriving conditions for the case where oxygen gradients dominate. The second modelling approach was agent based and takes a microscopic view by simulating the movements of single molecules.

4.1. Oxygen diffusion at low-oxygen concentrations

Oxygen uptake by *E. coli* is extremely efficient. In a steady state below 100% AU, the oxygen is removed almost completely from the medium. This ability of an *E. coli* population in a chemostat is caused by two effects: (1) the

Michaelis constants (K_m) of the terminal oxidases are well below 1 μM (Table 2.1) allowing for a high-specific oxygen uptake rate even at low-oxygen concentrations and (2) the steady-state biomass increases with aerobiosis, leading to an increase of the absolute oxygen uptake rate until the dissolved oxygen concentration reaches the order of magnitude of the Michaelis constants.

Full aerobiosis (100% AU) is reached at a steady-state dissolved oxygen concentration of about 1 μM (Alexeeva, 2000). This corresponds to less than 1000 molecules of molecular oxygen in the volume of an *E. coli* cell (ca. 1 μm^3 according to Kubitschek et al., 1986). According to Alexeeva et al. (2000), the specific oxygen uptake flux at full aerobiosis (≥100% AU) is about 6 mmol/g/h. Assuming a cell dry weight of 150 fg (BNID 103894; Milo, Jorgensen, Moran, Weber, & Springer, 2010), this is equivalent to a flux of about 150,000 oxygen molecules per second per cell. This analysis suggests that oxygen is cleared rapidly near the membrane. The question arises if oxygen gradients near and in cells constrain the oxygen response, for example, if the cytoplasm remains anaerobic even if the immediate environment of the membrane is aerobic. This hypothesis is supported by the observation that the cytoplasmatic TF FNR whose activity is directly inhibited by intracellular oxygen is active under microaerobic conditions (Section 4.1.2) (Rolfe et al., 2012).

4.2. A reaction–diffusion model of oxygen near and in *E. coli* cells

Unden and Schirawski (1997) and Unden et al. (1995) compute the oxygen diffusion rate into *E. coli* cells and conclude that oxygen diffusion could only be limiting for FNR signalling at oxygen concentrations below 0.2 mbar. Becker, Holighaus, Gabrielczyk, and Unden (1996) used an oxystat to analyse the effect of oxygen on FNR-dependent gene expression. They conclude that respiratory protection where oxygen-sensitive cytosolic enzymes are protected by rapid respiration does not play a role for FNR-dependent gene expression. However they consider oxygen concentrations in the order of 1–5 mbar that are higher than the oxygen concentrations at 100% AU. What stays is the question of the spatial oxygen profiles in and near cells at low-oxygen concentrations as they are attained at 100% AU and below.

The system under consideration is shown in Fig. 2.6. Oxygen diffuses to the cell membrane, where it is either consumed by the terminal oxidases or diffuses further into the cell interior. Several processes are able to remove

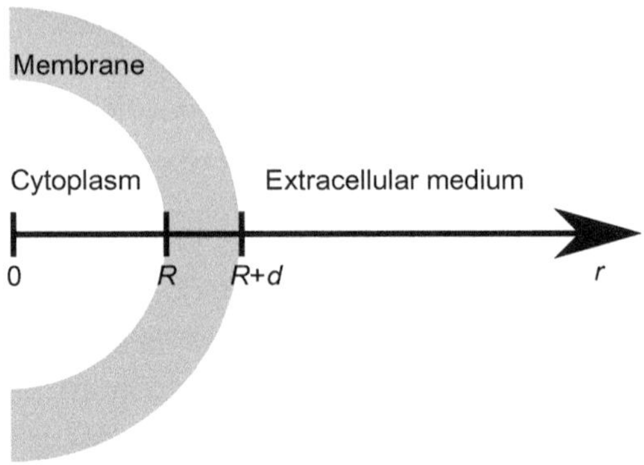

Figure 2.6 Spatial structure of the reaction–diffusion model. The coordinate axis *r* shows the distance from the cell centre.

oxygen from the cell interior. The inactivation of FNR consumes oxygen that is not released upon its reactivation. Further, soluble oxidases or oxygenases may contribute to cytoplasmic oxygen consumption, but this is probably minimal.

For the sake of simplicity, a spherical cell and spherical symmetry of the oxygen concentration is assumed. Then, the only spatial coordinate that we need to consider is the distance *r* from the centre of the cell. The partial-differential reaction–diffusion equation based on Fick's law in spherical coordinates is

$$\frac{\partial c}{\partial t}(t,r) = D\frac{\partial^2 c}{\partial r^2}(t,r) + D\frac{2}{r}\frac{\partial c}{\partial r}(t,r) - v(t,r)$$

with time t (s) and the distance r (m) (Crank, 1979). The equation describes the spatial and temporal behaviour of the concentration of oxygen c (M) in relation to the diffusion coefficient D (m^2/s) and the flow of oxygen consumption $v(t,r)$ (M/s).

The characteristic time scale of diffusion is given by $\tau \approx l^2/D$. Consider the surrounding of an *E. coli* cell ($l \approx 10$ μm) and the diffusion coefficient of oxygen in water ($D = 2.1 \times 10^{-9}$ m^2/s), the characteristic time is only $\tau \approx 0.5$ ms. Focusing on the biologically relevant time scales of metabolic and genetic regulation, we can assume that oxygen diffusion is in a steady state with $\partial c/\partial t(t,r) = 0$.

The above equation for the concentration c can be rewritten into an equation for the concentration difference $\Delta c(r) = c(r) - c_0$ between the concentration at radius r and the concentration c_0 at radius 0:

$$0 = D\frac{\partial^2 \Delta c}{\partial r^2}(r) + D\frac{2}{r}\frac{\partial \Delta c}{\partial r}(r) - v(r)$$

with the boundary condition $\Delta c(0)=0$. Due to the spherical symmetry, no oxygen diffusion occurs over the centre of the cell and we get the second boundary condition $\partial \Delta c/\partial r(t,0)=0$. The total oxygen consumption of the cell is given by J_{tot} (mol/s). By J_{cyt} (mol/s) and J_{mem} (mol/s), we denote the oxygen consumption in the cytoplasm and at the membrane, respectively. The parameter ξ determines the fraction of oxygen consumed in the membrane: $J_{cyt}=(1-\xi)J_{tot}$ and $J_{mem}=\xi J_{tot}$. The volume-specific flux $v(t,r)$ (M/s) is adjusted such that the oxygen consumption rates J_{cyt} and J_{mem} (mol/s) are realized in the cytoplasm and the membrane, respectively.

Figure 2.7 shows the calculated profile of the steady-state concentration difference Δc over r for different values of ξ and the parameter values shown in Table 2.3. The model does not account for the unequal equilibrium partition of oxygen between a lipid phase and an aqueous solution. Concentrations of oxygen in the membrane are given as the according equilibrium concentration in an aqueous solution. The diffusion coefficient in the membrane is corrected accordingly (see comment in Table 2.3).

The largest concentration difference occurs if ξ is small that is if most of the oxygen is consumed in the cytoplasm and not at the membrane. However, it is a justifiable assumption that most of the oxygen consumption occurs at the membrane-bound ETC (see Section 2.3) and thus ξ is expected to be near 1; henceforth, a value for $\xi=0.75$ is assumed. The prediction of this model is that an overall concentration difference of around 0.02 μM is realistic where about a quarter is an intracellular and the rest an extracellular difference. The concentration difference between the two sides of the

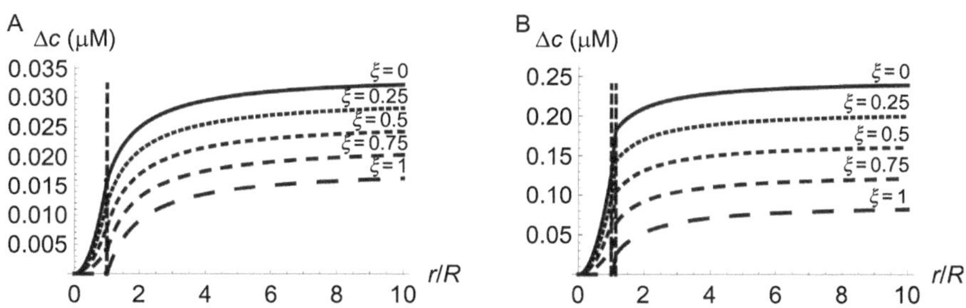

Figure 2.7 Concentration difference Δc over r for two different parameter sets. (A) Nominal parameter set from Table 2.3. (B) Disturbed parameter set $D_{cyt}=0.5D^*_{cyt}$, $D_{mem}=0.1D^*_{mem}$, $J_{tot}=2J^*_{tot}$, $d=10\cdot d^*$.

Table 2.3 Nominal parameters of the reaction–diffusion model

Parameter	Value	Unit	Sources	Remark
D^*_{cyt}	1.15×10^{-9}	m^2/s	Unden et al. (1995)	
D^*_{mem}	32.0×10^{-9}	m^2/s	Fischkoff and Vanderkooi (1975)	a,b
D^*_{ex}	2.00×10^{-9}	m^2/s	BNID 10440 Milo et al. (2010) and Kreft, Picioreanu, Wimpenny, and van Loosdrecht (2001)	Water
J^*_{tot}	1.50×10^5	Molecules/s	Alexeeva et al. (2000)	c
R^*	5.43×10^{-7}	m	BNID 100011 Milo et al. (2010)	d
d^*	3.75×10^{-9}	m	BNID 105823 Milo et al. (2010)	

[a]Coefficient for erythrocyte plasma membrane.
[b]Diffusion coefficient for an oxygen partition membrane/water of 1. Unequal oxygen partition between membrane and water is not incorporated into the model equations. Oxygen concentration in the membrane is underestimated by a factor of around 4.4. This is compensated by an increase of the diffusion coefficient by a factor of 4.4 (see Fischkoff & Vanderkooi, 1975).
[c]Calculated from a value 6 mmol/g/h assuming a dry cell mass of 150 fg (BNID 103894; Milo et al., 2010).
[d]Calculated from the cytoplasm volume assuming a spherical geometry.

membrane is negligible, which is in accordance with previous results (Becker et al., 1996; Unden et al., 1995; Unden & Schirawski, 1997).

The parameters in Table 2.3 are subject to some uncertainties. In particular, the diffusion constants in cytoplasm and membrane are critical parameters whose values are experimentally assessed in artificial situations. Also, the model neglects further possible diffusion barriers as the periplasm and the outer membrane. So, it is possible that the diffusion coefficients are overestimated and that the width of the membrane might need to be increased to include more than just the plasma membrane. The value of J_{tot} depends on the culture conditions and is growth rate dependent. Figure 2.8 shows the concentration differences between the cell centre and the inner side of the membrane $\Delta c(R)$, the outer side of the membrane $\Delta c(R+d)$ and the medium far away from the cell $\Delta c(\infty)$ for a variation of the model parameters away from their nominal value. For moderate variations in one parameter (factor 10 in the diffusion coefficients and factor 4 in the geometrical parameters and J_{tot}), a total concentration difference in the order of 0.1 μM is predicted. These effects superimpose for a simultaneous variation of parameters. Figure 2.7B shows the profile of Δc over r for a parameter vector, where the uncertain parameters are varied simultaneously. For $\xi = 0.75$, a total concentration difference of 0.1 μM is reached wherefrom a considerable part occurs in the membrane region.

As discussed above, any oxygen gradients that are formed will be dependent on the values of diffusion coefficients, geometric parameters and the

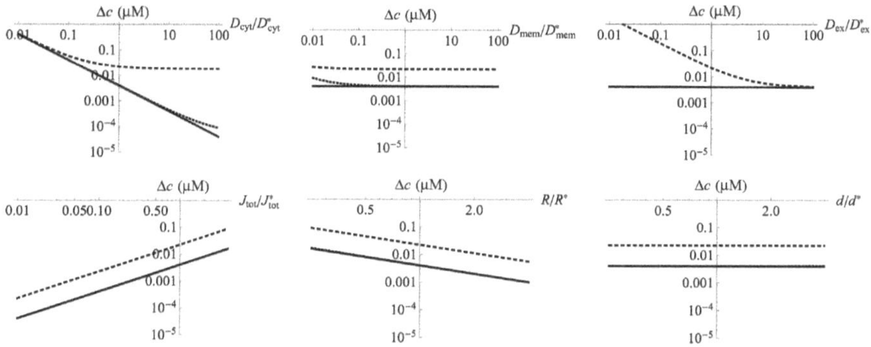

Figure 2.8 Dependency of the concentration differences for $\xi = 0.75$ on deviations from the nominal parameter values in Table 2.3. The first row shows the dependency on the diffusion coefficients of cytoplasm, membrane and extracellular medium (D_{cyt}, D_{mem} and D_{ex}). The second row shows the dependency on the total respiratory rate J_{tot}, the radius of the cell R and the thickness of the membrane d. concentration difference between the cell centre and the inner side of the membrane $\Delta c(R)$. Blue Dotted: concentration difference between the cell centre and the outer side of the membrane $\Delta c(R+d)$. Dashed: concentration difference between a point far away from the cell and the cell centre $\Delta c(\infty)$.

total oxygen uptake rate. For a set of nominal parameters (Table 2.3), a concentration difference of 0.02 μM is obtained. For a parameter set with impaired diffusion, a value of 0.1 μM is predicted. It is still an open question whether these calculated values support the hypothesis that near- and intracellular oxygen gradients play a pivotal role in the oxygen response at and below 100% AU. For a growth rate of 0.15 h^{-1}, Alexeeva (2000) reports an average oxygen concentration of about 1 μM in the region from 60% to 100% AU and a value of about 0.25 μM between 10% and 40% AU. This suggests that the oxygen gradients may be in the order of 1–10% of the oxygen concentration. It is important to note that at these low concentrations, oxygen electrodes work at their detection limit and measurements may not be very accurate. For example, Rolfe et al. (2012) reported that no oxygen was detectable up to 80% AU. This means that the actual ratio of oxygen gradient to concentration may be relevant at low aerobiosis values. Therefore, a closer, more microscopic view on oxygen diffusion and consumption is necessary and this is considered in the next section.

4.3. Agent-based model of FNR activity

To gain further insight into the potential influence of spatial constraints on the responses of FNR to changes in oxygen availability, an agent-based model of the regulation of terminal oxidase synthesis was developed.

Agent-based modelling is a relatively new way to analyse biological systems (Coakley, Smallwood, & Holcombe, 2006; Holcombe et al., 2012; Pogson, Holcombe, Smallwood, & Qwarnstrom, 2008, Pogson, Smallwood, Qwarnstrom, & Holcombe, 2006). It attempts to encapsulate the fact that much of biology is driven by local interactions and its immediate geometry and it is difficult to model this very precisely when using traditional approaches such as differential equations or stochastic processes—a cell is not a bag of chemicals moving around randomly, there is complex structure involved in most processes. These spatial constraints can be captured in great detail using agent-based modelling. Therefore, a realistic geometric space was created to represent a typical *E. coli* cell, within which, features such as membranes, cytoplasm and nucleoid were located. Into this structure, a population of the key molecules, each represented as autonomous agents, was introduced. Thus, realistic (based on experimental measurements) numbers of agents representing the terminal oxidase proteins, cytochrome *bd-I* (Cyd agent) and cytochrome *bo′* (Cyo agent), and the sensor, ArcB, were located in the inner membrane, whilst the regulators FNR and ArcA were placed in the cytoplasm and nucleoid. Molecules of oxygen were positioned outside the cell, where they are generated by the simulation to represent the arrival of oxygen in the experimental set-up. Each molecular agent is then programmed to behave as it would in reality. The agents will interact with other appropriate agents under suitable circumstances. For example, an oxygen agent will combine with a Cyo agent if they passes close enough to each other, as a consequence H_2O will be produced; an FNR dimer will bind to an unoccupied DNA-binding site, if it is close enough to do so. The agents move in various ways, some respecting boundaries such as membrane surfaces, others constrained by the biological information available. These behaviours are coded up in the FLAME language. The FLAME system automatically generates a simulation programme in the computer language C which can be run on a supercomputer—or a desktop running any operating system. To begin a simulation, it is necessary to define a starting position for all the agents that describes where they are located in the model and what state they are in—for example, whether they are free or bound to another agent. A convenient starting point is the anaerobic steady state (0% AU) and the simulation begins with the supply of external oxygen molecules. The simulation outputs the detailed locations, state and other information about every agent at every time step.

Two forms of model were constructed to explore the mechanism of the response of *E. coli* to different oxygen availabilities. The first included both

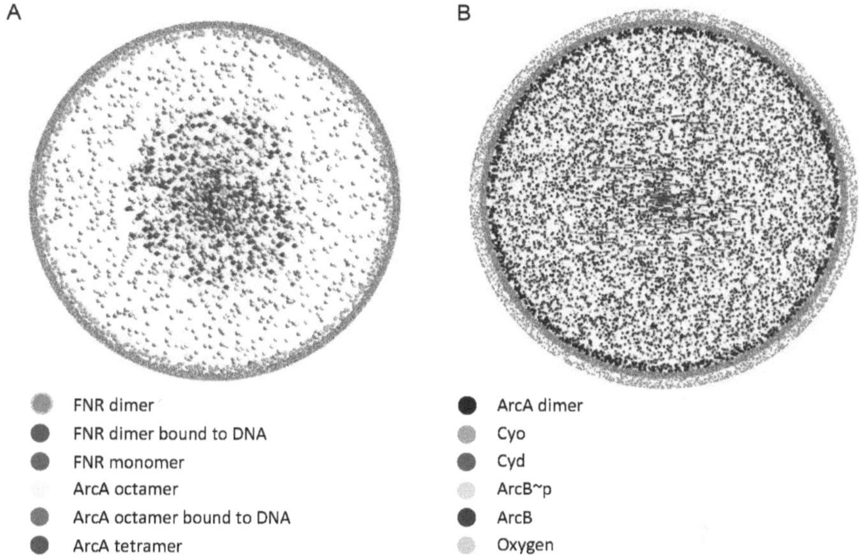

Figure 2.9 The initial and final states of model with no oxygen and with sufficient oxygen supplied. (A) The initial state of model with no oxygen supplied to the cell. (B) The final state of model with sufficient oxygen (217% AU) supplied to the cell. Each type of molecule is represented by a different grey level as shown. (See the color plate.)

ArcBA and FNR and was used to investigate the response of the two regulators when oxygen was supplied to the cell. The initial (0% AU) and final (217% AU) states of the cell are shown in Fig. 2.9. In the initial state (Fig. 2.9A), ArcB, ArcA and FNR are active and the DNA-binding sites for ArcA ~ P and FNR in the nucleoid are fully occupied. In the final state (217% AU), active FNR dimers are disassembled to monomers and ArcA is dephosphorylated causing these regulators to leave their cognate-binding sites in the nucleoid (Fig. 2.9B). During transition from 0% to 217% AU, it was noted that oxygen consumption by the terminal oxidases at the membrane and reaction of oxygen with FNR dimers in the cytoplasm served to protect the FNR dimers bound at target promoters in the nucleoid. This 'respiratory protection' could explain why FNR itself appears to have no in-built buffering capacity to prevent rapid switching between the 'on' and 'off' states.

The second form of the model was a simplified FNR-only model. Experimental data at 31% AU were used to fit the model and to define relative parameters, such as interaction radii. The obtained parameters were then used for other AU levels to validate the model against experimental data

Table 2.4 Comparison of the measured response of FNR to oxygen availability to the output of agent-based modelling

	Simulation results (average value)		Experimental results
Aerobiosis units (%)	Total FNR dimer (molecules per cell)	FNR dimer bound to binding sites (molecules per cell)	Total FNR dimer (molecules per cell)[a]
0	1500	350	1500
31	1350	350	1377
85	1033	287	1020
115	280	7	248
217	73	0	60

The model simulation data were taken by averaging the model outputs at steady state.
[a]The total numbers of FNR dimers were calculated from the Western blots and FNR-dependent promoter activities.

and the results are shown in Table 2.4. These outputs matched the FNR activities calculated from measurements of transcription from an FNR-dependent reporter (Table 2.4) and thus demonstrate the ability of the model to simulate the general behaviour of FNR in steady-state cultures of *E. coli*.

The agent-based model developed in SUMO recognizes the importance of geometry and location in biology. Highly detailed information, such as number, position and status, of any molecule can be monitored whilst the model runs. This offers the possibility of developing a deeper understanding of these adaptive mechanisms. For example, the prediction that there are two barriers to oxygen-mediated inactivation of FNR-dependent transcription, oxygen consumption by oxidases at the membrane and oxygen reaction with free FNR dimers in the cytoplasm accounts for the predicted (TFInfer) and observed activity of FNR, thereby providing new insight into the molecular mechanisms of the FNR switch, and generates new hypotheses for experimental investigation.

5. CONCLUSIONS

E. coli remains arguably the most intensively studied of all organisms. Furthermore, its metabolism and respiration were amongst the earliest aspects of its physiology to be investigated and were well represented in microbiology texts more than six decades ago. Molecular genetic, biochemical, physiological and structural studies have together created a vast reservoir of knowledge in the intervening years. Despite this, a comprehensive understanding of the holistic operation of the respiratory apparatus, that is a systems level comprehension, is lacking. The international consortium SUMO has begun to tackle this problem and the main findings are summarized here. Not only have these studies advanced our knowledge base but also, perhaps equally important, identified gaps in our knowledge so that future studies may be better directed.

When oxygen is available, *E. coli* uses respiration for redox balancing and optimal energy conservation via the PMF. The switch between fermentative and aerobic conditions and also between the states and transients induced by different levels of oxygen availability is driven by extensive changes in gene expression and protein synthesis and turnover, resulting in widespread changes in metabolic fluxes and concentrations. This oxygen response is determined by the interaction of global and local genetic regulatory mechanisms as well as enzymatic regulation, but constrained by physical parameters such as diffusion, thermodynamics and the requirement for a balance of carbon, electrons and energy (predominantly the PMF and ATP pools). Thus, a comprehensive systems level understanding of the bacterial oxygen response requires data acquired on different temporal and spatial scales (and necessarily of widely differing quality) hand-in-hand with the interpretation and integration of these data and the construction and testing of mathematical models. The data acquisition phases of such an undertaking are better advanced than the modelling, yet the inadequacy of some datasets limits the development of more refined models. Nevertheless, our approach has revealed a much more detailed picture of the aerobe–anaerobe response especially for the environmentally critical microaerobic range poised between unlimited oxygen availability and anaerobiosis.

The intermediate oxygen tensions that are encountered in natural niches are particularly under studied. In the SUMO work, a key feature has been a consortium-wide attempt to define and control precisely the oxygen status of chemostat cultures, not by direct oxygen measurements, but by determining the minimum oxygen concentration at which zero acetate is detectable, which is then defined as 100% aerobiosis. As oxygen provision decreases, the progressive increase in acetate flux, which reaches a maximum under anoxic conditions, is a physiological measure of oxygen availability for the respiratory system and TFs. In our work, the simplest transcriptomic approach measured transcript profiles of steady-state cultures, complemented by studying transitions from oxygen-replete to oxygen-starved conditions and vice versa. Thus, it was possible to capture complementary transcriptomic datasets that could be integrated with other biochemical and physiological measurements. The TfInfer analysis, which assumes a log-linear approximation for the transcriptomic response, revealed oxygen-dependent changes during successive steady states in 23 TFs including seven so-called global regulators related to energetic state. More surprisingly, the activity of the oxygen-sensing TF, FNR was predicted and experimentally shown to decrease rapidly only at aerobiosis values greater than 56%, whereas the activity of the indirect oxygen-sensor ArcBA was predicted to decrease linearly with increasing aerobiosis. Because ArcA phosphorylation is inhibited by oxidized quinone, we expected the quinone pool to be increasingly oxidized as the level of aerobiosis increased, but this was not the case. The relationship between the redox state of the quinone pool and ArcBA activity is not as simple as was initially envisaged and this, as well as the contributions of all the quinone species to both reductant flux and ArcBA regulation, and the role of allosteric effects of fermentation products, is an area that requires further study. Transitions between oxygen-replete and oxygen-deficient environments were used to mimic real-life environmental conditions allowing adaptation to new oxygen concentrations to be experimentally investigated, initially using transcriptomics. Here, the abundances of many more transcripts change during transitions than were observed in steady-state cultures, presumably reflecting the dynamic nature of the adaptive processes. Once metabolism has been reprogrammed, these acute stresses no longer exert their influence on the transcriptome and are thus not apparent from analysis of steady-state cultures. The inference of TF activities in dynamic environments also provided a means of unpicking the complex regulatory events that occur during adaptation. A comparison of the responses of TF activities when aerobic or anaerobic steady states were perturbed revealed a core set of 16 oxygen-responsive TFs (including FNR and ArcA) The network contains not only TFs that respond to oxygen availability but also regulate central metabolism, metal-ion homeostasis, oxidative stress responses, the nucleoid and amino acid biosynthesis.

A reaction–diffusion model of oxygen near and in *E. coli* cells predicted an overall oxygen concentration difference of around 0.02 μM across the membrane. Overall, the SUMO work suggests that, even on the scale of the bacterial cell, spatial distributions constrain physiology, for example by allowing 'hybrid' metabolism to proceed under microaerobic conditions during which anaerobic processes can be supported in the cytoplasm, whilst aerobic respiration proceeds in the vicinity of the cell membrane. To gain further insight into the potential influence of spatial constraints on the responses of FNR to changes in oxygen availability, an agent-based model of the regulation of terminal oxidase synthesis was developed, recognizing the importance of geometry and location in cells. The three terminal oxidases have distinct relationships with oxygen: cytochrome *bd-I* has an oxygen affinity significantly higher than that of cytochromes *bd-II* and *bo'*. The prediction that there are barriers to oxygen-mediated inactivation of FNR-dependent transcription, that is, oxygen consumption by the oxidases at the membrane and oxygen reaction with free FNR dimers in the cytoplasm now generates new hypotheses for experimental investigation.

Our steady-state model simulation of the bacterial cell at different values of aerobiosis allows qualitative and quantitative features of steady-state flux distributions to be described. Flux distributions are largely determined by thermodynamic constraints and justify a phenomenological but thermodynamically consistent approach for modelling the metabolism. In some cases, the measured metabolite curves are not reproduced well, suggesting either missing regulatory interactions or measurement error. These too represent areas for further study. However, gene expression can, in many cases, be reproduced realistically suggesting that the major genetic regulatory systems relevant for the oxygen response of the wild type are known. TF activities for ArcA and FNR are in agreement with the model output.

Thus, a true quantitative systems level understanding of the oxygen response of *E. coli* is not yet reached and will require the correct description not only of the wild type but also of mutant strains. The present model provides a first step towards a thermodynamically consistent view of the interactions between metabolites, enzymes and TFs of *E. coli* in a chemostat.

ACKNOWLEDGEMENTS

We would like to thank all members of the SysMO SUMO consortium for fruitful discussions and collaborations especially M. Bekker, F. Bruggemann, B. Cseke, E.D. Gilles, A. Graham, S. Henkel, W. Jia, A. Maleki-Dizaji, T. Nye, S. Stagge, T. Sauter, J. Teixeira de Mattos, A. Ter Beek. We thank the ERA Sysbio SysMO (Systems Biology of Microorganisms) initiative for funding the SUMO and $SUMO_2$ consortia. The research was funded by the Biotechnology and Biological Sciences Research Council (BBSRC), Bundesministerium für Bildung und Forschung (BMBF) and the Nederlandse Organisatie voor Wetenschappelijk Onderzoek (NWO).

REFERENCES

Abramson, J., Larsson, G., Byrne, B., Puustinen, A., Garcia-Horsman, A., & Iwata, S. (2000). Purification, crystallization and preliminary crystallographic studies of an integral membrane protein, cytochrome bo3 ubiquinol oxidase from Escherichia coli. *Acta Crystallographica Section D, Biological Crystallography, 56*, 1076–1078.

Abramson, J., Riistama, S., Larsson, G., Jasaitis, A., Svensson-Ek, M., Laakkonen, L., et al. (2000). The structure of the ubiquinol oxidase from Escherichia coli and its ubiquinone binding site. *Nature Structural Biology*, 7, 910–917.

Alberty, R. A. (2003). *Thermodynamics of biochemical reactions*. Hoboken, NJ, USA: John Wiley and Sons, Inc.

Alberty, R. A. (2006). Biochemical thermodynamics: Applications of Mathematica. *Methods of Biochemical Analysis, 48*, 1–458.

Alexeeva, S. V. (2000). *Molecular physiology of responses to oxygen in* Escherichia coli*: The role of the ArcAB system*. Universiteit van Amsterdam.

Alexeeva, S., Hellingwerf, K. J., & Teixeira de Mattos, M. J. (2002). Quantitative assessment of oxygen availability: Perceived aerobiosis and its effect on flux distribution in the respiratory chain of *Escherichia coli*. *Journal of Bacteriology, 184*, 1402–1406.

Alexeeva, S., Hellingwerf, K. J., & Teixeira de Mattos, M. J. (2003). Requirement of ArcA for redox regulation in Escherichia coli under microaerobic but not anaerobic or aerobic conditions. *Journal of Bacteriology, 185*, 204–209.

Alexeeva, S., Kort, B. D. E., Sawers, G., Klaas, J., Mattos, M. J. T. De, Hellingwerf, K. J., et al. (2000). Effects of limited aeration and of the ArcAB system on intermediary pyruvate catabolism in *Escherichia coli*. *Journal of Bacteriology, 182*, 4934–4940.

Alvarez, A. F., & Georgellis, D. (2010). In vitro and in vivo analysis of the ArcB/A redox signaling pathway. *Methods in Enzymology, 471*, 205–228.

Alvarez, A. F., Rodriguez, C., & Georgellis, D. (2013). Ubiquinone and menaquinone electron carriers represent the Yin and Yang in the redox regulation of the ArcB sensor kinase. *Journal of Bacteriology, 195*, 3054–3061.

Anraku, Y. (1988). Bacterial electron transport chains. *Annual Review of Biochemistry, 57*, 101–132.

Arras, T., Schirawski, J., & Unden, G. (1998). Availability of O_2 as a substrate in the cytoplasm of bacteria under aerobic and microaerobic conditions. *Journal of Bacteriology, 180*, 2133–2136.

Ashcroft, J. R., & Haddock, B. A. (1975). Synthesis of alternative membrane-bound redox carriers during aerobic growth of Escherichia coli in the presence of potassium cyanide. *The Biochemical Journal, 148*, 349–352.

Asif, H. M. S., Rolfe, M. D., Green, J., Lawrence, N. D., Rattray, M., & Sanguinetti, G. (2010). TFInfer: A tool for probabilistic inference of transcription factor activities. *Bioinformatics, 26*, 2635–2636.

Beard, D. A. (2005). A biophysical model of the mitochondrial respiratory system and oxidative phosphorylation. *PLoS Computational Biology, 1*, e36.

Becker, S., Holighaus, G., Gabrielczyk, T., & Unden, G. (1996). O_2 as the regulatory signal for FNR-dependent gene regulation in Escherichia coli. *Journal of Bacteriology, 178*, 4515–4521.

Bekker, M., Alexeeva, S., Laan, W., Sawers, G., Teixeira de Mattos, J., & Hellingwerf, K. (2010). The ArcBA two-component system of Escherichia coli is regulated by the redox state of both the ubiquinone and the menaquinone pool. *Journal of Bacteriology, 192*, 746–754.

Bekker, M., de Vries, S., Ter Beek, A., Hellingwerf, K. J., & de Mattos, M. J. T. (2009). Respiration of Escherichia coli can be fully uncoupled via the nonelectrogenic terminal cytochrome bd-II oxidase. *Journal of Bacteriology, 191*, 5510–5517.

Belevich, I., Borisov, V. B., Konstantinov, A. A., & Verkhovsky, M. I. (2005). Oxygenated complex of cytochrome bd from Escherichia coli: Stability and photolability. *FEBS Letters, 579*, 4567–4570.

Bergersen, F. J., & Turner, G. L. (1979). Systems utilizing oxygenated leghemoglobin and myoglobin as sources of free dissolved O_2 at low concentrations for experiments with bacteria. *Analytical Biochemistry, 96*, 165–174.

Borisov, V., Arutyunyan, A. M., Osborne, J. P., Gennis, R. B., & Konstantinov, A. A. (1999). Magnetic circular dichroism used to examine the interaction of *Escherichia coli* cytochrome *bd* with ligands. *Biochemistry, 38*, 740–750.

Borisov, V. B., Belevich, I., Bloch, D. A., Mogi, T., & Verkhovsky, M. I. (2008). Glutamate 107 in subunit I of cytochrome bd from Escherichia coli is part of a transmembrane intraprotein pathway conducting protons from the cytoplasm to the heme b595/heme d active site. *Biochemistry, 47*, 7907–7914.

Borisov, V. B., Gennis, R. B., Hemp, J., & Verkhovsky, M. I. (2011). The cytochrome bd respiratory oxygen reductases. *Biochimica et Biophysica Acta, 1807*, 1398–1413.

Borisov, V. B., Murali, R., Verkhovskaya, M. L., Bloch, D. A., Han, H., Gennis, R. B., et al. (2011). Aerobic respiratory chain of Escherichia coli is not allowed to work in fully uncoupled mode. *Proceedings of the National Academy of Sciences of the United States of America, 108*, 17320–17324.

Brown, T. D. K., Jones-Mortimer, M. C., & Kornberg, H. L. (1977). The enzymic interconversion of acetate and acetyl-coenzyme A in *Escherichia coli*. *Journal of General Microbiology, 102*, 327–336.

Calhoun, M. W., Oden, K. L., Gennis, R. B., de Mattos, M. J., & Neijssel, O. M. (1993). Energetic efficiency of Escherichia coli: Effects of mutations in components of the aerobic respiratory chain. *Journal of Bacteriology, 175*, 3020–3025.

Cecchini, G., Schröder, I., Gunsalus, R. P., & Maklashina, E. (2002). Succinate dehydrogenase and fumarate reductase from *Escherichia coli*. *Biochimica et Biophysica Acta, 1553*, 140–157.

Clark, D. (1989). The fermentation pathways of *Escherichia coli*. *FEMS Microbiology Letters, 63*, 223–234.

Coakley, S., Smallwood, R., & Holcombe, M. (2006). From molecules to insect communities—How formal agent based computational modelling is uncovering new biological facts. *Scientiae Mathematicae Japonicae Online, e-2006*, 765–778.

Constantinidou, C., Hobman, J. L., Griffiths, L., Patel, M. D., Penn, C. W., Cole, J. A., et al. (2006). A reassessment of the FNR regulon and transcriptomic analysis of the effects of nitrate, nitrite, NarXL, and NarQP as Escherichia coli K12 adapts from aerobic to anaerobic growth. *The Journal of Biological Chemistry, 281*, 4802–4815.

Crack, J. C., Green, J., Hutchings, M. I., Thomson, A. J., & Le Brun, N. E. (2012). Bacterial iron-sulfur regulatory proteins as biological sensor-switches. *Antioxidants and Redox Signaling, 17*, 1215–1231.

Crank, J. (1979). *The mathematics of diffusion*. Ely House, London W.1, UK: Oxford University Press.

D'Mello, R., Hill, S., & Poole, R. K. (1994). Determination of the oxygen affinities of terminal oxidases in Azotobacter vinelandii using the deoxygenation of oxyleghaemoglobin and oxymyoglobin: Cytochrome bd is a low-affinity oxidase. *Microbiology, 140*, 1395–1402.

D'mello, R., Hill, S., & Poole, R. K. (1995). The oxygen affinity of cytochrome bo′ in Escherichia coli determined by the deoxygenation of oxyleghemoglobin and oxymyoglobin: Km values for oxygen are in the submicromolar range. *Journal of Bacteriology, 177*, 867–870.

D'Mello, R., Hill, S., & Poole, R. K. (1996). The cytochrome bd quinol oxidase in Escherichia coli has an extremely high oxygen affinity and two oxygen-binding haems:

Implications for regulation of activity in vivo by oxygen inhibition. *Microbiology*, *142*, 755–763.

Dassa, J., Fsihi, H., Marck, C., Dion, M., Kieffer-Bontemps, M., & Boquet, P. L. (1991). A new oxygen-regulated operon in Escherichia coli comprises the genes for a putative third cytochrome oxidase and for pH 2.5 acid phosphatase (*appA*). *Molecular and General Genetics*, *229*, 341–352.

Davidge, K. S., Sanguinetti, G., Yee, C. H., Cox, A. G., McLeod, C. W., Monk, C. E., et al. (2009). Carbon monoxide-releasing antibacterial molecules target respiration and global transcriptional regulators. *The Journal of Biological Chemistry*, *284*, 4516–4524.

Du Plessis, D. J. F., Nouwen, N., & Driessen, A. J. M. (2006). Subunit a of cytochrome o oxidase requires both YidC and SecYEG for membrane insertion. *The Journal of Biological Chemistry*, *281*, 12248–12252.

Ederer, M. (2010). *Thermokinetic modeling and model reduction of reaction networks*. Dissertation, Germany: University of Stuttgart.

Ederer, M., & Gilles, E. D. (2007). Thermodynamically feasible kinetic models of reaction networks. *Biophysical Journal*, *92*, 1846–1857.

El-Mansi, E. M. T., & Holms, W. H. (1989). Control of carbon flux to acetate excretion during growth of *Escherichia coli* in batch and continuous cultures. *Microbiology*, *135*, 2875–2883.

Fischkoff, S., & Vanderkooi, J. M. (1975). Oxygen diffusion in biological and artificial membranes determined by the fluorochrome pyrene. *The Journal of General Physiology*, *65*, 663–676.

Georgellis, D. (1998). Signal decay through a reverse phospho relay in the Arc two-component signal transduction system. *The Journal of Biological Chemistry*, *273*, 32864–32869.

Georgellis, D. (1999). Amplification of signaling activity of the Arc two-component system of Escherichia coli by anaerobic metabolites. An in vitro study with different protein modules. *The Journal of Biological Chemistry*, *274*, 35950–35954.

Georgellis, D., Kwon, O., & Lin, E. C. (2001). Quinones as the redox signal for the arc two-component system of bacteria. *Science*, *292*, 2314–2316.

Georgellis, D., Lynch, A., & Lin, E. (1997). In vitro phosphorylation study of the Arc two-component signal transduction system of *Escherichia coli*. *Journal of Bacteriology*, *179*, 5429–5435.

Green, J., Scott, C., & Guest, J. R. (2001). Functional versatility in the CRP-FNR superfamily of transcription factors: FNR and FLP. *Advances in Microbial Physiology*, *44*, 1–34.

Holcombe, M., Adra, S., Bicak, M., Chin, S., Coakley, S., Graham, A. I., et al. (2012). Modelling complex biological systems using an agent-based approach. *Integrative Biology*, *4*, 53–64.

Holms, H. (1996). Flux analysis and control of the central metabolic pathways in *Escherichia coli*. *FEMS Microbiology Reviews*, *19*, 85–116.

Hoskisson, P. A., & Hobbs, G. (2005). Continuous culture—Making a comeback? *Microbiology*, *151*, 3153–3159.

Ingledew, W. J., & Poole, R. K. (1984). The respiratory chains of *Escherichia coli*. *Microbiological Reviews*, *48*, 222–271.

Iuchi, S., & Lin, E. C. (1992a). Purification and phosphorylation of the Arc regulatory components of *Escherichia coli*. *Journal of Bacteriology*, *174*, 5617–5623.

Iuchi, S., & Lin, E. C. (1992b). Mutational analysis of signal transduction by ArcB, a membrane sensor protein responsible for anaerobic repression of operons involved in the central aerobic pathways in *Escherichia coli*. *Journal of Bacteriology*, *174*, 3972–3980.

Jeon, Y., Lee, Y. S., Han, J. S., Kim, J. B., & Hwang, D. S. (2001). Multimerization of phosphorylated and non-phosphorylated ArcA is necessary for the response regulator function of the Arc two-component signal transduction system. *The Journal of Biological Chemistry, 276*, 40873–40879.

Jesse, H. E., Nye, T. L., McLean, S., Green, J., Mann, B. E., & Poole, R. K. (2013). Cytochrome bd-I in Escherichia coli is less sensitive than cytochromes bd-II or bo″ to inhibition by the carbon monoxide-releasing molecule, CORM-3: N-acetylcysteine reduces CO-RM uptake and inhibition of respiration. *Biochimica et Biophysica Acta, 1834*, 1693–1703.

Kang, Y., Weber, K. D., Qiu, Y., Kiley, P. J., & Blattner, F. R. (2005). Genome-wide expression analysis indicates that FNR of Escherichia coli K-12 regulates a large number of genes of unknown function. *Journal of Bacteriology, 187*, 1135–1160.

Kaysser, T. M., Ghaim, J. B., Georgiou, C., & Gennis, R. B. (1995). Methionine-393 is an axial ligand of the heme b558 component of the cytochrome bd ubiquinol oxidase from *Escherichia coli*. *Biochemistry, 34*, 13491–13501.

Keseler, I. M., Mackie, A., Peralta-Gil, M., Santos-Zavaleta, A., Gama-Castro, S., Bonavides-Martínez, C., et al. (2013). EcoCyc: Fusing model organism databases with systems biology. *Nucleic Acids Research, 41*, D605–D612.

Kettner, C., & Hicks, M. (2005). The dilemma of modern functional enzymology. *Current Enzyme Inhibition, 1*, 171–181.

Klamt, S., Grammel, H., Straube, R., Ghosh, R., & Gilles, E. D. (2008). Modeling the electron transport chain of purple non-sulfur bacteria. *Molecular Systems Biology, 4*, 156.

Kreft, J.-U., Picioreanu, C., Wimpenny, J. W. T., & van Loosdrecht, M. C. M. (2001). Individual-based modelling of biofilms. *Microbiology, 147*, 2897–2912.

Kubitschek, H. E., Friske, J. A., Counter-analyzer, C., Trueba, F. S., England, N., & Instruments, B. (1986). Determination of bacterial cell volume with the Coulter Counter. *Journal of Bacteriology, 168*, 1466–1467.

Landwall, P., & Holme, T. (1977). Influence of glucose and dissolved oxygen concentrations on yields of *Escherichia coli* B in dialysis culture. *Journal of General Microbiology, 103*, 353–358.

Lindqvist, A., Membrillo-Hernández, J., Poole, R. K., & Cook, G. M. (2000). Roles of respiratory oxidases in protecting Escherichia coli K12 from oxidative stress. *Antonie Van Leeuwenhoek, 78*, 23–31.

Liu, X., & De Wulf, P. (2004). Probing the ArcA-P modulon of Escherichia coli by whole genome transcriptional analysis and sequence recognition profiling. *The Journal of Biological Chemistry, 279*, 12588–12597.

Maklashina, E., Berthold, D. A., & Cecchini, G. (1998). Anaerobic expression of *Escherichia coli* succinate dehydrogenase: Functional replacement of fumarate reductase in the respiratory chain during anaerobic growth. *Journal of Bacteriology, 180*, 5989–5996.

Malpica, R., Franco, B., Rodriguez, C., Kwon, O., & Georgellis, D. (2004). Identification of a quinone-sensitive redox switch in the ArcB sensor kinase. *Proceedings of the National Academy of Sciences of the United States of America, 101*, 13318–13323.

Malpica, R., Sandoval, G. R. P., Rodríguez, C., Franco, B., & Georgellis, D. (2006). Signaling by the arc two-component system provides a link between the redox state of the quinone pool and gene expression. *Antioxidants and Redox Signaling, 8*, 781–795.

Martínez-Antonio, A., Salgado, H., Gama-Castro, S., Gutiérrez-Ríos, R. M., Jiménez-Jacinto, V., & Collado-Vides, J. (2003). Environmental conditions and transcriptional regulation in Escherichia coli: A physiological integrative approach. *Biotechnology and Bioengineering, 84*, 743–749.

Mason, M. G., Shepherd, M., Nicholls, P., Dobbin, P. S., Dodsworth, K. S., Poole, R. K., et al. (2009). Cytochrome bd confers nitric oxide resistance to Escherichia coli. *Nature Chemical Biology, 5*, 94–96.

Mills, C. E., Sedelnikova, S., Søballe, B., Hughes, M. N., & Poole, R. K. (2001). Escherichia coli flavohaemoglobin (Hmp) with equistoichiometric FAD and haem contents has a low affinity for dioxygen in the absence or presence of nitric oxide. *The Biochemical Journal, 353*, 207–213.

Milo, R., Jorgensen, P., Moran, U., Weber, G., & Springer, M. (2010). BioNumbers—The database of key numbers in molecular and cell biology. *Nucleic Acids Research, 38*, D750–D753.

Mogi, T., Akimoto, S., Endou, S., Watanabe-Nakayama, T., Mizuochi-Asai, E., & Miyoshi, H. (2006). Probing the ubiquinol-binding site in cytochrome bd by site-directed mutagenesis. *Biochemistry, 45*, 7924–7930.

Moore, A. L., Carré, J. E., Affourtit, C., Albury, M. S., Crichton, P. G., Kita, K., et al. (2008). Compelling EPR evidence that the alternative oxidase is a diiron carboxylate protein. *Biochimica et Biophysica Acta, 1777*, 327–330.

Nicholls, D. G., & Ferguson, S. J. (2013). *Bioenergetics* (4th ed.). London, NW1, 7BY, UK: Academic Press.

Park, S.-J., Tseng, C.-P., & Gunsalus, R. P. (1995). Regulation of succinate dehydrogenase (*sdhCDAB*) operon expression in *Escherichia coli* in response to carbon supply and anaerobiosis: Role of ArcA and Fnr. *Molecular Microbiology, 15*, 473–482.

Partridge, J. D., Scott, C., Tang, Y., Poole, R. K., & Green, J. (2006). *Escherichia coli* transcriptome dynamics during the transition from anaerobic to aerobic conditions. *The Journal of Biological Chemistry, 281*, 27806–27815.

Picon, A., Teixeira de Mattos, M. J., & Postma, P. W. (2005). Reducing the glucose uptake rate in Escherichia coli affects growth rate but not protein production. *Biotechnology and Bioengineering, 90*, 191–200.

Pirt, S. J. (1985). *Principles of microbe and cell cultivation*. Oxford, UK: Blackwell Publishing Ltd.

Pogson, M., Holcombe, M., Smallwood, R., & Qwarnstrom, E. (2008). Introducing spatial information into predictive NF-kappaB modelling—An agent-based approach. *PLoS One, 3*, e2367.

Pogson, M., Smallwood, R., Qwarnstrom, E., & Holcombe, M. (2006). Formal agent-based modelling of intracellular chemical interactions. *Biosystems, 85*, 37–45.

Poole, R. K., Anjum, M. F., Membrillo-Hernández, J., Kim, S. O., Hughes, M. N., & Stewart, V. (1996). Nitric oxide, nitrite, and Fnr regulation of hmp (flavohemoglobin) gene expression in *Escherichia coli* K-12. *Journal of Bacteriology, 178*, 5487–5492.

Poole, R. K., & Chance, B. (1995). Paper commentary. *Microbiology, 141*, 752–753.

Poole, R. K., & Cook, G. M. (2000). Redundancy of aerobic respiratory chains in bacteria? Routes, reasons and regulation. *Advances in Microbial Physiology, 43*, 165–224.

Poole, R. K., & Hill, S. (1997). Respiratory protection of nitrogenase activity in Azotobacter vinelandii—Roles of the terminal oxidases. *Bioscience Reports, 17*, 303–317.

Poole, R. K., & Hughes, M. N. (2000). New functions for the ancient globin family: Bacterial responses to nitric oxide and nitrosative stress. *Molecular Microbiology, 36*, 775–783.

Poole, R. K., Lloyd, D., & Chance, B. (1979). The reaction of cytochrome oxidase with oxygen in the fission yeast *Schizosaccharomyces pombe* 972 h. Studies at subzero temperatures and measurement of apparent oxygen affinity. *The Biochemical Journal, 184*, 555–563.

Poole, R. K., Salmon, I., & Chance, B. (1983). The reaction with oxygen of cytochrome oxidase (cytochrome d) in *Escherichia coli* K12: Optical studies of intermediate species and cytochrome b oxidation at sub-zero temperatures. *Journal of General Microbiology, 129*, 1345–1355.

Poole, R. K., & Williams, H. D. (1987). Proposal that the function of the membrane-bound cytochrome a1-like haemoprotein (cytochrome b-595) in *Escherichia coli* is a direct electron donation to cytochrome d. *FEBS Letters, 217*, 49–52.

Poole, R. K., Williams, H. D., Downie, J. A., & Gibson, F. (1989). Mutations affecting the cytochrome d-containing oxidase complex of Escherichia coli K12: Identification and mapping of a fourth locus, cydD. *Microbiology, 135*, 1865–1874.

Puustinen, A., Finel, M., Haltia, T., Gennis, R. B., & Wikström, M. (1991). Properties of the two terminal oxidases of *Escherichia coli*. *Biochemistry, 30*, 3936–3942.

Puustinen, A., Finel, M., Virkki, M., & Wikström, M. (1989). Cytochrome o (bo) is a proton pump in *Paracoccus denitrificans* and *Escherichia coli*. *FEBS Letters, 249*, 163–167.

Reed, J., Vo, T., Schilling, C., & Palsson, B. (2003). An expanded genome-scale model of Escherichia coli K-12 (iJR904 GSM/GPR). *Genome Biology, 4*, R54.

Rodriguez, C., Kwon, O., & Georgellis, D. (2004). Effect of D-lactate on the physiological activity of the ArcB sensor kinase in Escherichia coli. *Journal of Bacteriology, 186*, 2085–2090.

Rolfe, M. D., Beek, A. Ter, Graham, A. I., Trotter, E. W., Asif, H. M. S., Sanguinetti, G., et al. (2011). Transcript profiling and inference of *Escherichia coli* K-12 ArcA activity across the range of physiologically relevant oxygen concentrations. *The Journal of Biological Chemistry, 286*, 10147–10154.

Rolfe, M. D., Ocone, A., Stapleton, M. R., Hall, S., Trotter, E. W., Robert, K., et al. (2012). Systems analysis of transcription factor activities in environments with stable and dynamic oxygen concentrations. *Open Biology, 2*, 120091.

Saiki, K., Nakamura, H., Mogi, T., & Anraku, Y. (1996). Probing a role of subunit IV of the Escherichia coli bo-type ubiquinol oxidase by deletion and cross-linking analyses. *The Journal of Biological Chemistry, 271*, 15336–15340.

Salgado, H., Peralta-Gil, M., Gama-Castro, S., Santos-Zavaleta, A., Muñiz-Rascado, L., García-Sotelo, J. S., et al. (2013). RegulonDB v8.0: Omics data sets, evolutionary conservation, regulatory phrases, cross-validated gold standards and more. *Nucleic Acids Research, 41*, D203–D213.

Salmon, K., Hung, S., Mekjian, K., Baldi, P., Hatfield, G. W., & Gunsalus, R. P. (2003). Global gene expression profiling in Escherichia coli K12. The effects of oxygen availability and FNR. *The Journal of Biological Chemistry, 278*, 29837–29855.

Salmon, K. A., Hung, S., Steffen, N. R., Krupp, R., Baldi, P., Hatfield, G. W., et al. (2005). Global gene expression profiling in Escherichia coli K12: Effects of oxygen availability and ArcA. *The Journal of Biological Chemistry, 280*, 15084–15096.

Sanguinetti, G., Lawrence, N. D., & Rattray, M. (2006). Probabilistic inference of transcription factor concentrations and gene-specific regulatory activities. *Bioinformatics, 22*, 2775–2781.

Schoepp-Cothenet, B., Lieutaud, C., Baymann, F., Verméglio, A., Friedrich, T., Kramer, D. M., et al. (2009). Menaquinone as pool quinone in a purple bacterium. *Proceedings of the National Academy of Sciences of the United States of America, 106*, 8549–8554.

Sharma, P. (2013). *On the efficiency of catabolism at the cellular level*. PhD thesis, University of Amsterdam.

Sharma, P., Hellingwerf, K. J., de Mattos, M. J. T., & Bekker, M. (2012). Uncoupling of substrate-level phosphorylation in Escherichia coli during glucose-limited growth. *Applied and Environmental Microbiology, 78*, 6908–6913.

Sharma, P., Stagge, S., Bekker, M., Bettenbrock, K., & Hellingwerf, K. J. (2013). Kinase activity of ArcB from Escherichia coli is subject to regulation by both ubiquinone and demethylmenaquinone. *PLoS One, 8*, e75412.

Shen, J., & Gunsalus, R. P. (1997). Role of multiple ArcA recognition sites in anaerobic regulation of succinate dehydrogenase (sdhCDAB) gene expression in *Escherichia coli*. *Molecular Microbiology, 26*, 223–236.

Shepherd, M., Sanguinetti, G., Cook, G. M., & Poole, R. K. (2010). Compensations for diminished terminal oxidase activity in Escherichia coli: Cytochrome bd-II-mediated respiration and glutamate metabolism. *The Journal of Biological Chemistry, 285,* 18464–18472.

Slonczewski, J. L., & Foster, J. W. (2008). *Microbiology: An evolving science.* W.W. Norton & Company, Inc.

Smith, H. L., & Waltman, P. (2008). *The theory of the chemostat: Dynamics of microbial competition (Cambridge studies in mathematical biology)* (2nd ed.). Cambridge University Press.

Søballe, B., & Poole, R. K. (1999). Microbial ubiquinones: Multiple roles in respiration, gene regulation and oxidative stress management. *Microbiology, 145*(Pt. 8), 1817–1830.

Steinsiek, S., Frixel, S., Stagge, S., & Bettenbrock, K. (2011). Characterization of E. coli MG1655 and frdA and sdhC mutants at various aerobiosis levels. *Journal of Biotechnology, 154,* 35–45.

Stenberg, F., von Heijne, G., & Daley, D. O. (2007). Assembly of the cytochrome bo3 complex. *Journal of Molecular Biology, 371,* 765–773.

Sturr, M. G., Krulwich, T. A., & Hicks, D. B. (1996). Purification of a cytochrome bd terminal oxidase encoded by the Escherichia coli app locus from a Δcyo Δcyd strain complemented by genes from *Bacillus firmus* OF4. *Journal of Bacteriology, 178,* 1742–1749.

Taylor, B. L., & Zhulin, I. B. (1999). PAS domains: Internal sensors of oxygen, redox potential, and light. *Microbiology and Molecular Biology Reviews, 63,* 479–506.

Trotter, E. W., Rolfe, M. D., Hounslow, A. M., Craven, C. J., Williamson, M. P., Sanguinetti, G., et al. (2011). Reprogramming of Escherichia coli K-12 metabolism during the initial phase of transition from an anaerobic to a micro-aerobic environment. *PLoS One, 6,* e25501.

Tseng, C., Albrecht, J., & Gunsalus, R. (1996). Effect of microaerophilic cell growth conditions on expression of the aerobic (cyoABCDE and cydAB) and anaerobic (narGHJI, frdABCD, and dmsABC) respiratory pathway genes in Escherichia coli. *Journal of Bacteriology, 178,* 1094–1098.

Unden, G., Becker, S., Bongaerts, J., Holighaus, G., Schirawski, J., & Six, S. (1995). O_2-sensing and O_2-dependent gene regulation in facultatively anaerobic bacteria. *Archives of Microbiology, 164,* 81–90.

Unden, G., & Bongaerts, J. (1997). Alternative respiratory pathways of Escherichia coli: Energetics and transcriptional regulation in response to electron acceptors. *Biochimica et Biophysica Acta, 1320,* 217–234.

Unden, G., & Schirawski, J. (1997). The oxygen-responsive transcriptional regulator FNR of Escherichia coli: The search for signals and reactions. *Molecular Microbiology, 25,* 205–210.

Visser, D., & Heijnen, J. J. (2003). Dynamic simulation and metabolic re-design of a branched pathway using linlog kinetics. *Metabolic Engineering, 5,* 164–176.

Wagner, A. M., Krab, K., Wagner, M. J., & Moore, A. L. (2008). Regulation of thermogenesis in flowering Araceae: The role of the alternative oxidase. *Biochimica et Biophysica Acta, 1777,* 993–1000.

Wikström, M., Krab, K., & Saraste, M. (1981). *Cytochrome oxidase—A synthesis.* London, New York: Academic Press.

Zhang, J., Barquera, B., & Gennis, R. B. (2004). Gene fusions with beta-lactamase show that subunit I of the cytochrome bd quinol oxidase from E. coli has nine transmembrane helices with the O_2 reactive site near the periplasmic surface. *FEBS Letters, 561,* 58–62.

Zhang, B., Crack, J. C., Subramanian, S., Green, J., Thomson, A. J., Le Brun, N. E., et al. (2012). Reversible cycling between cysteine persulfide-ligated [2Fe-2S] and cysteine-ligated [4Fe-4S] clusters in the FNR regulatory protein. *Proceedings of the National Academy of Sciences of the United States of America, 109,* 15734–15739.

CHAPTER THREE

The Silicon Trypanosome: A Test Case of Iterative Model Extension in Systems Biology

Fiona Achcar[*], Abeer Fadda[†], Jurgen R. Haanstra[‡,§], Eduard J. Kerkhoven[¶,||], Dong-Hyun Kim[¶], Alejandro E. Leroux[#], Theodore Papamarkou[**], Federico Rojas[††], Barbara M. Bakker[‡], Michael P. Barrett[¶], Christine Clayton[†], Mark Girolami[**], R. Luise Krauth-Siegel[#], Keith R. Matthews[††], Rainer Breitling[‡‡,1]

[*]Institute of Molecular, Cell and Systems Biology, College of Medical, Veterinary and Life Sciences, University of Glasgow, Glasgow, United Kingdom
[†]Zentrum für Molekulare Biologie der Universität Heidelberg, DKFZ-ZMBH Alliance, Heidelberg, Germany
[‡]Department of Pediatrics, Centre for Liver Digestive and Metabolic Diseases, and Systems Biology Centre for Energy Metabolism and Ageing, University Medical Center Groningen, University of Groningen, Groningen, The Netherlands
[§]Department of Molecular Cell Physiology, Faculty of Earth and Life Sciences, VU University Amsterdam, Amsterdam, The Netherlands
[¶]Wellcome Trust Centre for Molecular Parasitology, Institute of Infection, Immunity and Inflammation, and Glasgow Polyomics, College of Medical, Veterinary and Life Sciences, University of Glasgow, Glasgow, United Kingdom
[||]Systems and Synthetic Biology Group, Department of Chemical and Biological Engineering, Chalmers University of Technology, Gothenburg, Sweden
[#]Biochemie-Zentrum der Universität Heidelberg, Heidelberg, Germany
[**]The Department of Statistical Science and The Centre for Computational Statistics and Machine Learning University College London, London, United Kingdom
[††]Centre for Immunity, Infection and Evolution, Institute for Immunology and Infection Research, School of Biological Sciences, Ashworth Laboratories, University of Edinburgh, Edinburgh, United Kingdom
[‡‡]Manchester Institute of Biotechnology, Faculty of Life Sciences, University of Manchester, Manchester, United Kingdom
[1]Corresponding author: e-mail address: rainer.breitling@manchester.ac.uk

Contents

Advances in Microbial Physiology, Volume 64
ISSN 0065-2911
http://dx.doi.org/10.1016/B978-0-12-800143-1.00003-8

Abstract

The African trypanosome, *Trypanosoma brucei*, is a unicellular parasite causing African Trypanosomiasis (sleeping sickness in humans and nagana in animals). Due to some of its unique properties, it has emerged as a popular model organism in systems biology. A predictive quantitative model of glycolysis in the bloodstream form of the parasite has been constructed and updated several times. The Silicon Trypanosome is a project that brings together modellers and experimentalists to improve and extend this core model with new pathways and additional levels of regulation. These new extensions and analyses use computational methods that explicitly take different levels of uncertainty into account. During this project, numerous tools and techniques have been developed for this purpose, which can now be used for a wide range of different studies in systems biology.

1. INTRODUCTION

Trypanosoma brucei, a unicellular bloodstream parasite transmitted by tsetse flies, is the causative agent of African Trypanosomiasis (sleeping sickness in humans and nagana in animals; Barrett et al., 2003). In recent years, it has emerged as a popular model organism in systems biology (Barrett, Bakker, & Breitling, 2010), due to the wealth of information obtained in biomedical studies and the many advantages it offers: reproducible cultivation methods exist and quantitative analysis is well established. A quantitative mathematical model of central energy metabolism in the bloodstream form of the parasite was one of the first predictive kinetic models of a complex biological system (Bakker, Michels, Opperdoes, & Westerhoff, 1997), and this model has been iteratively updated after experimental testing and validation (Albert et al., 2005; Bakker et al., 2000; Bakker, Michels, Opperdoes, & Westerhoff, 1999; Haanstra, van Tuijl, et al., 2008; Helfert, Estévez, Bakker, Michels, & Clayton, 2001), yielding one of the most extensively curated kinetic metabolic models currently available.

The Silicon Trypanosome (SilicoTryp) is a project that brings together computational systems biologists and trypanosome experts to address multi-level regulation of trypanosome physiology on a new scale. Using a Bayesian approach that explicitly takes multiple levels of uncertainty into account, the project aims at re-analysing the core model of glycolysis and extends it in two directions: metabolic extensions towards new pathways ("horizontal" extensions) and inclusion of transcription and translation ("vertical" extensions). Explicitly including our uncertainty, both in the parameter values and the topologies of the different extensions, allows us not only to quantify our level of confidence in the results but also to determine, in an unbiased manner, pitfalls and inconsistencies that might have gone unnoticed if the modelling was done without uncertainty awareness.

The horizontal extensions initially focus on the pathways controlling redox balance in trypanosomes: the trypanothione pathway, which is critical for parasite survival in the host, and the pentose phosphate pathway that links the trypanothione pathway to glycolysis by providing the necessary NADPH. Adding these two pathways requires knowledge of the kinetic parameters of all enzymes involved. In order to obtain the most accurate parameter values for these additional enzymes, a new buffer was developed to mimic as closely as possible the conditions in the cytosol of the parasite, as has previously been done for yeast (van Eunen et al., 2010). The first enzyme characterised in this buffer was trypanothione synthetase, the final enzyme of the trypanothione biosynthesis pathway, using a method that integrates laboratory experimentation and modelling to explore alternative model topologies (Leroux, Haanstra, Bakker, & Krauth-Siegel, 2013).

In addition to these metabolic extensions, the SilicoTryp project aims at a vertical extension of the model by including additional levels of control: transcription and translation. In trypanosomes, transcription rates (with very few exceptions) play little role in the regulation of mRNA concentrations since transcription by RNA polymerase II is not regulated for individual genes (Siegel, Gunasekera, Cross, & Ochsenreiter, 2011). Instead, regulation is achieved at the levels of pre-mRNA- and mRNA-processing and degradation. These levels of control are modelled based on the previously published ordinary differential equations model of phosphoglycerate kinase (PGK) transcription and translation (Haanstra, Stewart, et al., 2008). The necessary parameters are acquired using deep sequencing (RNAseq). These estimates will be particularly uncertain, compared to metabolic enzyme parameters, and this provides an additional motivation for using a fully uncertainty-aware modelling strategy.

The uncertainty-aware modelling requires a rigorous method for updating our initial (prior) beliefs on the parameter values of the models, when new data becomes available. For this purpose, it was necessary to develop a computationally efficient Monte Carlo variance reduction method (Papamarkou, Mira, & Girolami, 2014). Two types of large-scale experimental data are generated within the SilicoTryp project to drive the iterative updating process: steady-state concentration of metabolites and metabolic fluxes were measured using liquid chromatography–mass spectrometry (LC–MS) in standard conditions, and a collection of RNAi mutants were constructed, to examine the behaviour of the system under a variety of perturbations. The mutants in this collection will be characterised at the metabolome, transcriptome and proteome level, providing new data to further refine the extended model and reduce our uncertainty on model topology and parameter values.

2. UNCERTAINTY IN THE GLYCOLYSIS MODEL

SilicoTryp explicitly deals with uncertainty at multiple levels (experimental data, kinetic parameter values, model topology) in every aspect of the project. It aims at applying the general framework of Fig. 3.1 (Vyshemirsky & Girolami, 2008; Xu et al., 2010).

The *T. brucei* glycolysis model, as previously published (Albert et al., 2005; Bakker et al., 2000, 1997; Bakker, Michels, et al., 1999; Haanstra, van Tuijl, et al., 2008; Helfert et al., 2001) (Fig. 3.2), used a single set of parameters values. Some basic uncertainty analysis was included by scanning a range of values for selected parameters individually (Bakker et al., 1997) or comparing two alternative topologies (Bakker, Michels, et al., 1999), but the combined effect of uncertainty in *all* parameters was not evaluated systematically, nor were the plausible distributions of parameter values explicitly defined. Therefore, an important step in the SilicoTryp project was to collect information not only on the (experimental and biological) uncertainty of all parameter values in the previously published model but also on the uncertainty of the topology of the metabolic network.

2.1. Parameter uncertainty in the glycolysis model

In order to evaluate uncertainty on the parameter values of the glycolysis model, a Mediawiki-based website was created, the SilicoTryp wiki (http://silicotryp.ibls.gla.ac.uk/wiki). In this wiki, the sources and calculations underlying the values for every parameter of the glycolysis model were documented (see e.g. in Fig. 3.3). This detailed source of documentation allowed us to define probability distributions for each parameter, hence

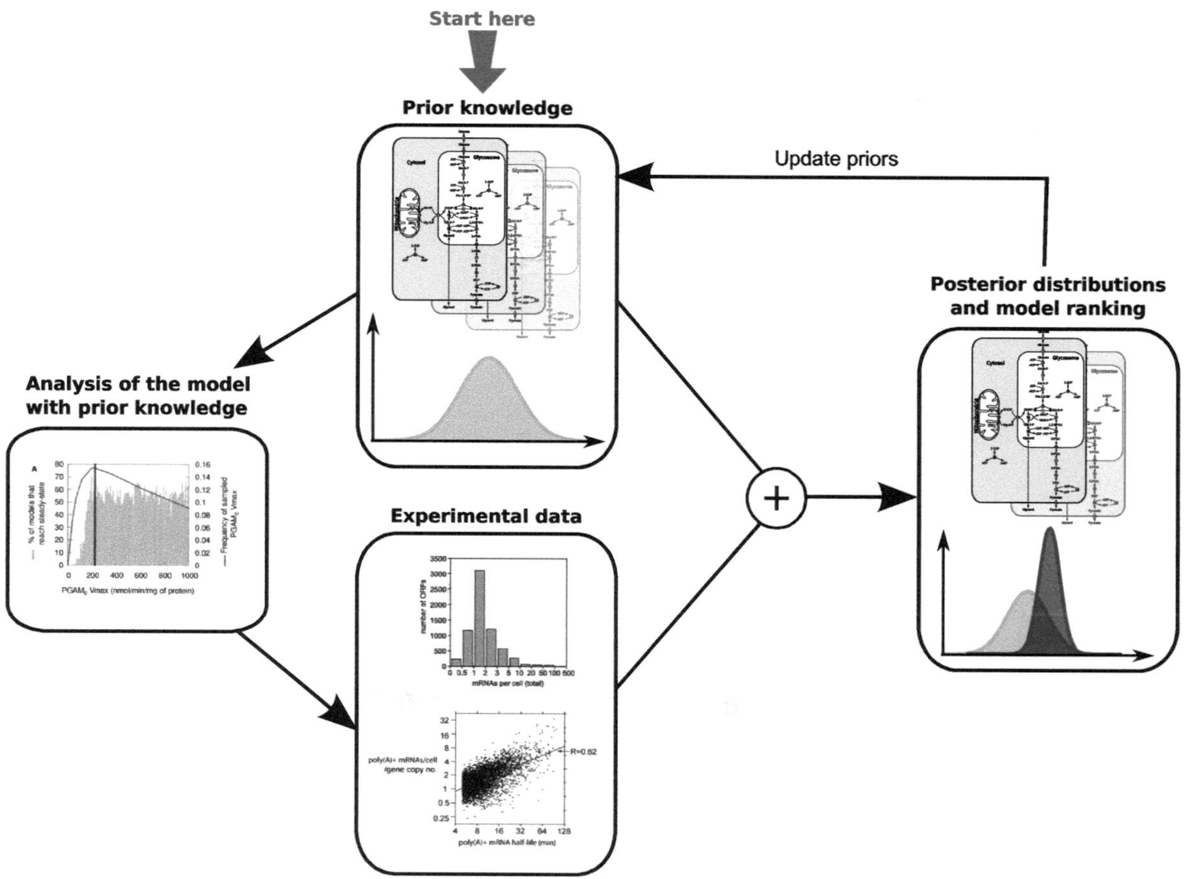

Figure 3.1 General Bayesian framework for explicitly handling uncertainty in the SilicoTryp project. (See the color plate.)

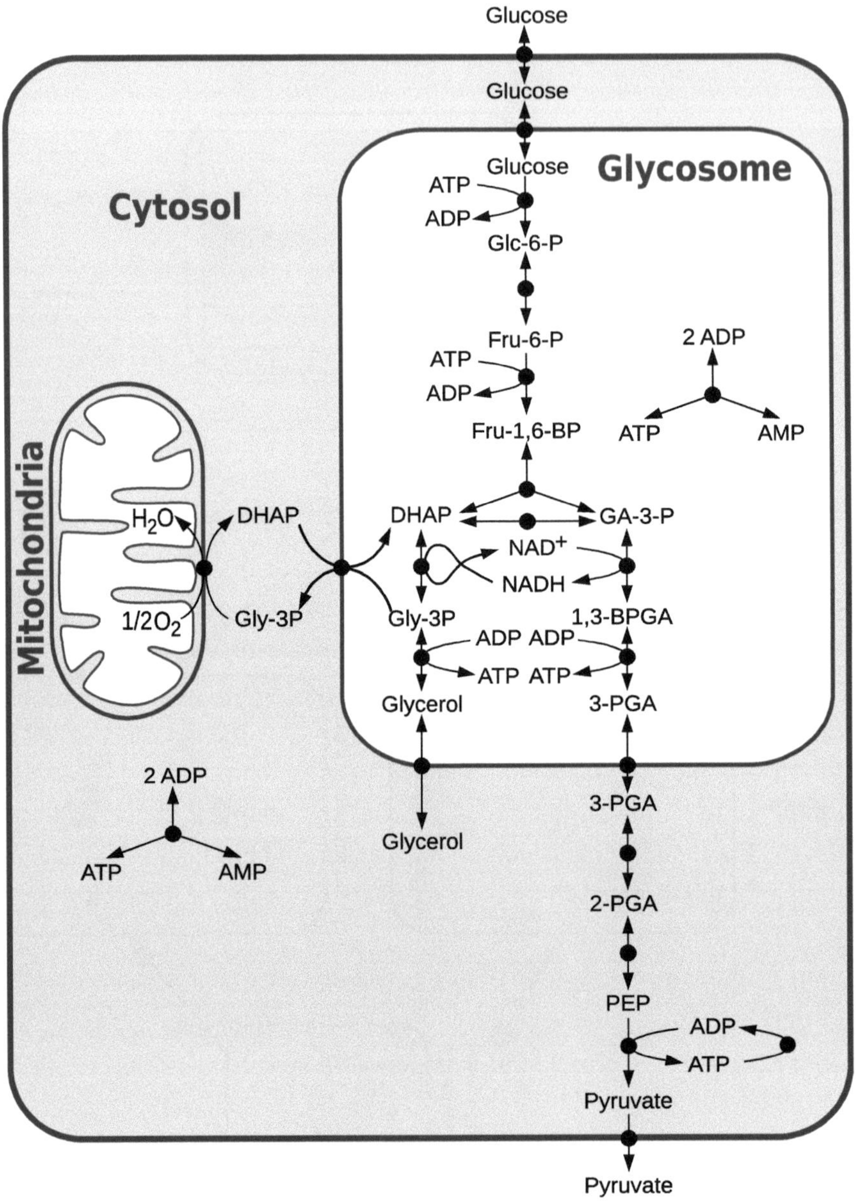

Figure 3.2 Glycolysis in T. *brucei* as previously published (Albert et al., 2005). *Abbreviations*: Glc-6-P, glucose 6-phosphate; Fru-6-P, fructose 6-phosphate; Fru-1,6-BP, fructose 1,6-bisphosphate; DHAP, dihydroxyacetone phosphate; GA-3-P, glyceraldehyde 3-phosphate; Gly-3-P, glycerol 3-phosphate; 1,3-BPGA, 1,3-bisphosphoglycerate; 3-PGA, 3-phosphoglycerate; 2-PGA, 2-phosphoglycerate; PEP, phosphoenolpyruvate. *Reprinted from Achcar et al. (2012), under the Creative Commons Attribution License.*

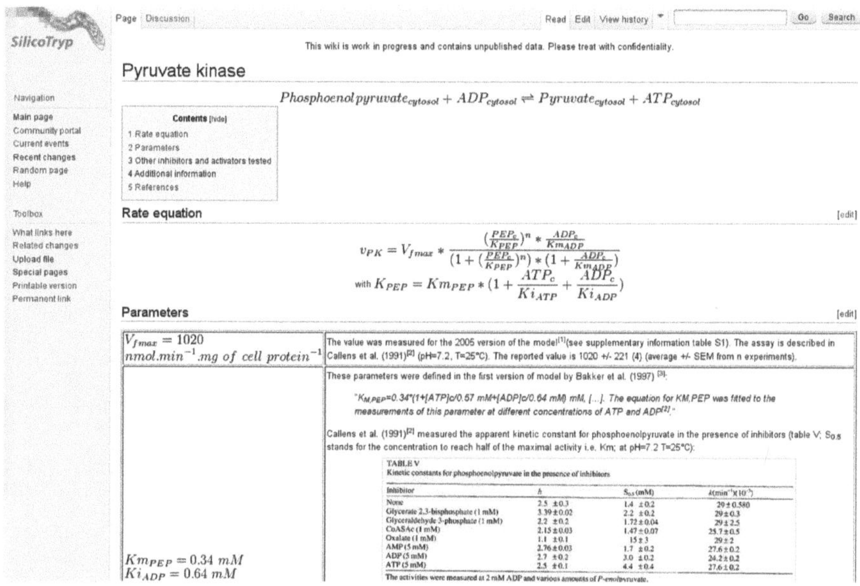

Figure 3.3 Example page of the SilicoTryp wiki. Chemical and rate equations are described, and sources for each parameter are documented in detail.

representing our prior beliefs on what the true parameter values are (Achcar et al., 2012).

Using these distributions, sets of plausible parameters were sampled. A large collection of variant models, each with a different set of sampled parameters, was then simulated, allowing us to quantify confidence intervals for different model properties, such as steady-state concentrations and fluxes or control coefficients resulting from the parameter uncertainty (see Fig. 3.4). This strategy allowed us not only to gain knowledge on how confident we are about the model predictions but also to detect fragilities of the model in an unbiased way, highlighting areas of the model behaviour that are in contradiction with experimental data (Achcar et al., 2012). We showed, for example, that only in 40% of our sampled models is the control over the glucose consumption flux mostly held by the glucose transporter, as described before with the fixed parameter model (Bakker, Michels, et al., 1999; Bakker, Walsh, et al., 1999). In the equally plausible scenarios represented by the remaining 60% of the models, this control is shared between several other glycolytic enzymes, and which enzymes share control over the glucose consumption flux depends on the parameter sets considered (Achcar et al., 2012).

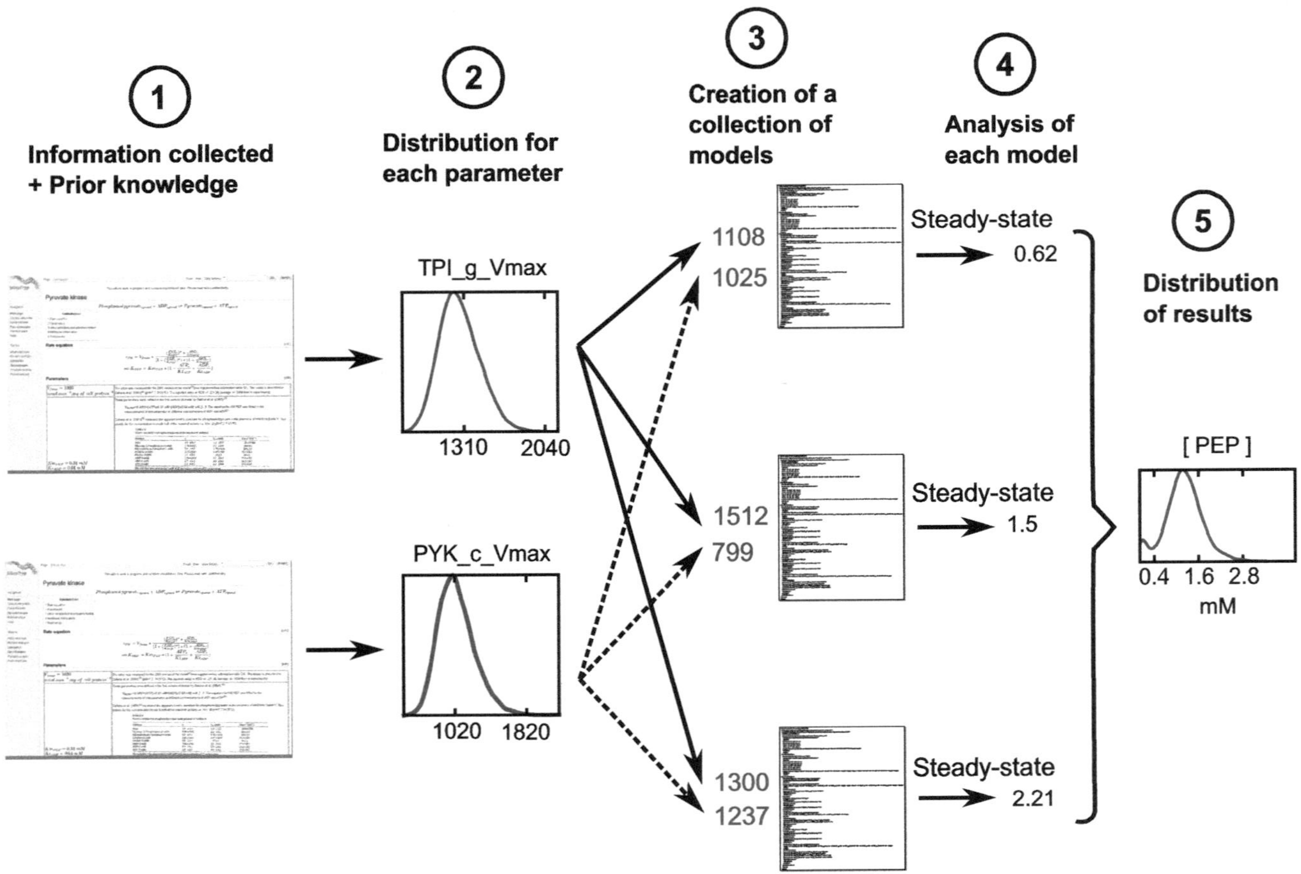

Figure 3.4 Evaluation of parameter uncertainties and their effect on the model results. *Reprinted with permission from Breitling, Achcar, and Takano (2013). Copyright 2013 American Chemical Society.*

2.2. Topological uncertainty in the glycolysis model

In addition to evaluating the effect of parameter value uncertainty, we explored the effect of topological uncertainty in the glycolysis model (Xu et al., 2010). Although *T. brucei* central energy metabolism has been exceptionally well studied for a long time, some areas of its network connectivity remain controversial. Indeed, the previously published model considered the glycosomes, organelles where most of the glycolytic enzymes are localised (see Fig. 3.2), as impermeable compartments into which all required metabolites need to be transported via specific transporters. This hypothesis was reinforced by both theoretical (Bakker et al., 2000) and experimental (Haanstra, van Tuijl, et al., 2008) findings that mislocalising the glycolytic enzymes to the cytosol causes a lethal accumulation of sugar phosphate. This confirmation of theoretical predictions by experimental results has been considered one of the major successes of the *T. brucei* glycolysis model.

However, recently, size-specific pores have been discovered in the glycosomal membrane (Gualdron-López et al., 2012). These pores are essentially chemically non-selective holes and their presence implies that small metabolites can diffuse freely across the membrane, while only the largest molecules would be retained inside the glycosomes. This creates additional potential links in the metabolic network, between cytosol and glycosome, which are not yet fully understood, but need to be considered in the metabolic modelling.

Another aspect of topological uncertainty is due to the fact that we know that enzymes are imported into the glycosome fully folded (Häusler, Stierhof, Wirtz, & Clayton, 1996; Michels et al., 2005); of course, these might be fully sequestered in transit by the protein import machinery, but there is also a concrete possibility that there could be some residual activity of glycosomal glycolytic enzymes in the cytosol. Cytosolic activity of the relevant enzymes has so far been poorly characterised and is generally neglected in modelling.

To test the effect of increasing permeability of the glycosome, a series of models was constructed (Achcar, Barrett, & Breitling, 2013). These models have increasingly permeable glycosomes, ranging from the impermeable glycosomes of the previous models (model 1a has a similar topology as the models published since 1999 (Bakker, Michels, et al., 1999), model 1b has a similar topology as the model of 1997 (Bakker et al., 1997)) to glycosomes permeable to all metabolites up to the size of ATP (model 6).

For each of these models, the uncertainty on the parameter values was taken into account as previously (Achcar et al., 2012). This was then extended by also including the uncertainty on the percentage of activity of the glycosomal enzymes in the cytosol. The resulting collection of models was simulated, and results were compared to experimental data (metabolite concentrations and fluxes) by computing log-likelihoods: the higher the log-likelihoods the better the match between simulations and experiments (see Fig. 3.5). Our results show that, quite surprisingly, the semi-permeable models that let metabolites smaller than fructose 6-phosphate (model 3) or smaller than fructose 1,6-bisphosphate (model 4) freely diffuse, agree with the experimental data at least as well as the impermeable model (model 1a). This is consistent with the idea that only compartmentation of ATP is required. Contrary to the common intuition that organelles like the glycosome function as tightly sealed compartments, at least for glycosomal glycolysis a certain degree of leakiness for smaller metabolites appears to be fully compatible with its function.

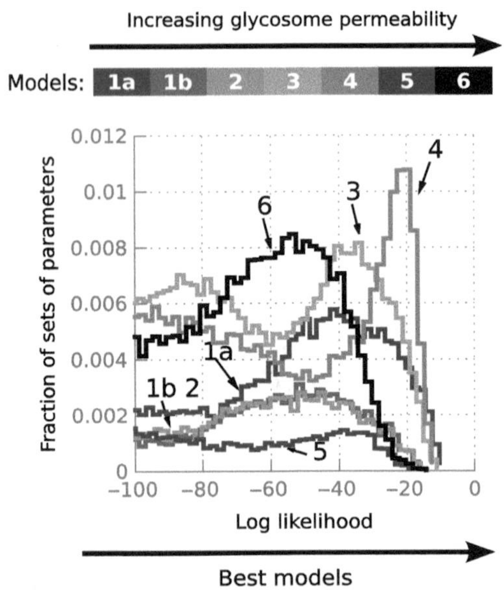

Figure 3.5 Log-likelihood (goodness of fit to experimental data) of models with increasing glycosome permeability and cytosolic activities of the glycosomal enzymes. The higher the log-likelihood, the better the match between simulations and experimental results. Each model is simulated with a range of parameter sets that describes our uncertainty about the parameter values. See Achcar et al. (2013) for details. *Modified with permission from Achcar et al. (2013). Copyright 2013 John Wiley and Sons.* (See the color plate.)

In addition, the analysis allowed us to show that, for all topologies with a semi-permeable glycosome tested, the fraction of residual cytosolic activity of some glycolytic enzymes probably plays a crucial role in explaining parasite physiology.

3. METABOLIC EXTENSIONS OF THE MODEL

With the uncertainty-aware model of central carbon metabolism established, the SilicoTryp project could aim at including the trypanothione and pentose phosphate pathways and the associated extra levels of uncertainty into the model.

3.1. The pentose phosphate pathway

In *T. brucei*, the pentose phosphate pathway is localised in both the cytosol and the glycosome where its presence breaks the bound-phosphate balance that was thought to be essential (Bakker et al., 2000; Haanstra, van Tuijl, et al., 2008). This phosphate "leak" needs to be solved either by finding a viable model that does not need bound-phosphate to be balanced in the glycosome, or by adding reactions to restore this balance. One solution within each of these two categories of solutions was tested (Kerkhoven et al., 2013).

The first solution tested maintained the bound-phosphate balance by adding ribokinase to the glycosome (hypothesis 1 of Fig. 3.6). In the context of impermeable glycosomes, ribokinase would work in the direction of producing ATP and ribose. To be able to model this solution, the enzyme was characterised biochemically (Kerkhoven et al., 2013). However, despite being sufficient theoretically, the results showed that this enzyme alone cannot maintain the bound-phosphate balance. The second solution modelled involves breaking the bound-phosphate balance by introducing an ATP/ADP translocator. However, modelling shows that this translocator would need to be very tightly regulated for the parasite to be viable. This makes it unlikely that such an antiporter alone could represent a solution and more reactions are probably involved in the regulation of the bound-phosphate balance.

Despite the topological uncertainty that remains, both of these model versions can already be used to some extent. Both of them handle oxidative stress in the same way, and the models have been used to test possible mechanisms to explain why the third enzyme of the PPP, 6-phosphogluconate dehydrogenase (from 6-PG to Rul-5-P in Fig. 3.6), is essential for parasite survival. The resulting predictions were then tested experimentally and it

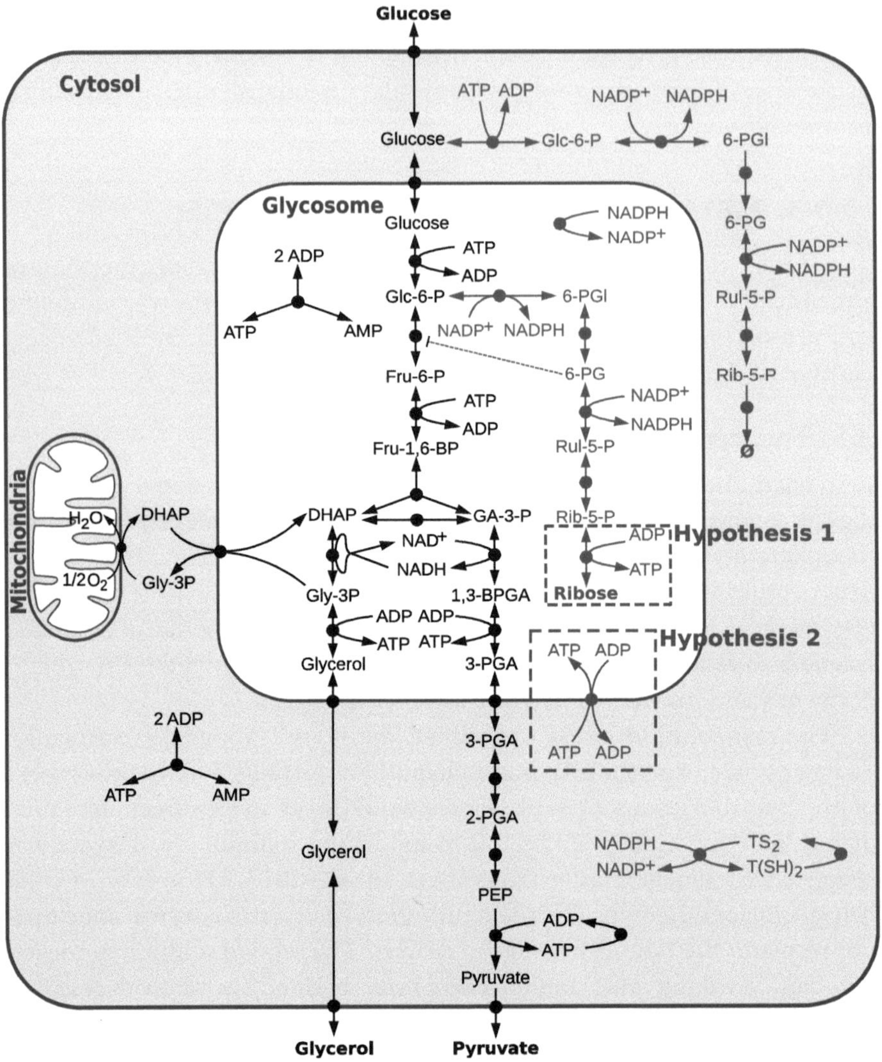

Figure 3.6 Two hypotheses tested to solve the topology of the pentose phosphate pathway. In black, the reactions of the glycolysis model, in grey the reactions introduced to model the pentose phosphate pathway. Abbreviations are not present in Fig. 3.2: 6-PGl, 6-phosphogluconolactone; 6-PG, 6-phosphogluconate; Rul-5-P, ribulose 5-phosphate; Rib-5-P, ribose 5-phosphate; TS_2, trypanothione disulfide; $T(SH)_2$, trypanothione.

was shown that a positive feedback loop related to 6-PG inhibition of phosphoglucose isomerase was not responsible for the 6-phosphogluconate dehydrogenase essentiality in *T. brucei* (Kerkhoven et al., 2013), in contrast to the situation in many other organisms.

3.2. The trypanothione pathway

3.2.1 The pathway topology

In order to extend the model with the trypanothione pathway, a main contributor to redox balance in *T. brucei*, we first had to determine the stoichiometric map of this pathway. We constructed a stoichiometric map of the trypanothione pathway based on the literature and LC–MS data (see Fig. 3.7). Cells were grown in the presence of 50% ^{13}C-labelled glucose over 48 h, after which their metabolites were extracted and analysed using LC–MS (Kim, D.H., Achcar, F. et al., in preparation). The partially labelled metabolome allowed us to rule out the presence of the methionine salvage pathway (marked 1 in Fig. 3.7) in *T. brucei*. Indeed, if this pathway—recycling methylthioadenosine (MTA) into methionine—was occurring, we should detect some four carbon-labelled methionine. However, methionine is clearly detected as fully unlabelled, indicating that the pathway is not complete (Kim, D.H., Achcar, F. et al., in preparation).

One question that remains open is the origin of ornithine *in vivo*. Previous data indicated that trypanosomes lack arginase (Vincent et al., 2012), the

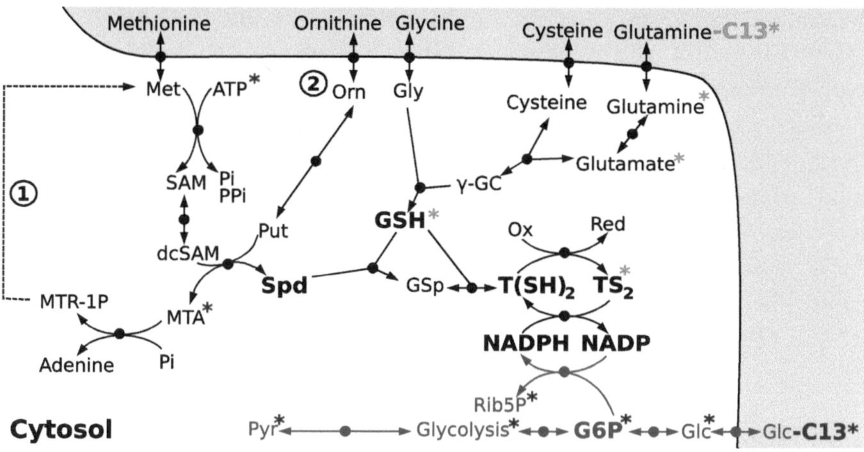

Figure 3.7 Stoichiometric map of the trypanothione pathway. The blue asterisks indicate which metabolites are found ^{13}C labelled when all glucose carbons are ^{13}C labelled. The turquoise asterisks indicate which metabolites are labelled when glutamine is labelled. The methionine salvage pathway, marked as 1, does not seem to be active in *T. brucei* (see text). The origin of ornithine, marked as 2, remains uncertain. *Abbreviations*: Orn, ornithine; Gly, glycine, Met, methionine; γ-GC, γ-glutamylcysteine; SAM, *S*-adenosylmethionine; dcSAM, decarboxylated *S*-adenosylmethionine; Spd, spermidine; Gsp, glutathionylspermidine; MTA, methylthioadenosine; MTR-1P, methylthioribose 1-phosphate; Glc, glucose; G6P, glucose 6-phosphate; Pyr, pyruvate; TS_2, trypanothione disulfide; $T(SH)_2$, trypanothione.

enzyme that produces ornithine from arginine, which is the common source of ornithine in eukaryotic cells including other parasites (Vincendeau, Gobert, Daulouède, Moynet, & Djavad Mossalayi, 2003). It was also shown that in the presence of ornithine, *T. brucei* is capable of using it directly, and therefore it was hypothesised that the cells use ornithine directly from the medium. However, measurements of the metabolite composition in the growth medium over time (Kim, D.H., Achcar, F. et al., in preparation; see below) showed that the ornithine concentration is quite low in the growth medium (Creek et al., 2013; see Section 5.2) and does not change over time (staying between 15 and 20 μM in the medium; it was measured as 54.4 ± 16.1 μM by Martens-Lobenhoffer et al. in blood (Martens-Lobenhoffer, Postel, Tröger, & Bode-Böger, 2007), i.e. the same order of magnitude). This suggests that there probably is another route to the production of ornithine in *T. brucei*, and its origin is currently being examined.

3.2.2 Uniform assay conditions

Usually, enzyme activities are measured under non-physiological optimum conditions. For a reliable in *silico* model and to reduce our uncertainty about the parameter values, the data should preferably be obtained under conditions that mimic the milieu in which the pathway is active in the cell (van Eunen et al., 2010). The importance of physiological conditions to measure metabolic functions was recently demonstrated: the use of an "*in vivo*-like" assay medium yielded enzyme kinetic parameters that substantially improved a computational model of yeast glycolysis (van Eunen, Kiewiet, Westerhoff, & Bakker, 2012). Such a buffer had not been used for *T. brucei* before. Based on the available literature, we therefore developed a new phosphate buffer system at pH 7.0 in which measurements are done at 37 °C (Table 3.1), mimicking the physiological environment of the enzymes in the cytosol of bloodstream-form parasites (Leroux et al., 2013).

3.2.3 A detailed focus on one enzyme: Trypanothione synthetase

We first employed our *in vivo*-like assay to characterise kinetically the enzyme of the final step in trypanothione ($T(SH)_2$) synthesis. *T. brucei* trypanothione synthetase (TryS) generates $T(SH)_2$ from glutathione (GSH) and spermidine (Spd) in two consecutive ATP-dependent reactions (Comini et al., 2004; Oza, Ariyanayagam, Aitcheson, & Fairlamb, 2003). In the first step, glutathionylspermidine (Gsp) is formed, which is then combined with a second GSH to yield $T(SH)_2$. *In vitro*, TryS also displays amidase activity and can, in total, catalyse five different reactions: ATP

Table 3.1 Characteristics of the *in vivo*-like buffer system for the cytosol of bloodstream *T. brucei*

Parameter	Value	Reference(s)
Temperature	37 °C	
pH	7.0	Nolan and Voorheis, (2000a) and Thissen and Wang (1991)
[Phosphate]	10 mM	Moreno et al. (2000)
[K^+]	100 mM	Nolan and Voorheis (2000b)
[Na^+]	15 mM	Nolan and Voorheis (2000b)
[Mg^{2+}]	10 mM	Fuad et al. (2011)
[Cl^-]	120 mM	Nolan and Voorheis (2000b)
Ionic strength	200 mM	

hydrolysis, formation of Gsp and $T(SH)_2$ as well as hydrolysis of the conjugates to regenerate GSH and Spd. As the mechanism of TryS is rather complex, a rate equation could not be easily formulated. To obtain a deeper insight in the exact molecular mechanism of the synthetase reactions of TryS, we formulated alternative kinetic models with different topologies of the catalytic cycle. Each elementary reaction step was modelled by a mass-action kinetic equation. The model parameters were then fitted by evolutionary programming to the extensive matrix of steady-state data obtained for different substrate/product combinations using our *in vivo*-like buffer. In the process of building the model, a strong interaction between experimental and modelling efforts proved to be essential. New enzyme kinetic experiments were designed to provide data to help discriminate between different models. Moreover, when the fit to the data was sub-optimal model topology was adjusted in an iterative cycle (see Fig. 3.8).

The best model describes the full kinetic profile of TryS and is also able to predict time profiles of (intermediate) product formation that were not used in the fitting of the parameters, and thus serve as validation data (Leroux et al., 2013). It also provides a mechanism for inhibition by the substrate glutathione and the product trypanothione. In our *in vivo*-like assay, some of the kinetic constants proved markedly different from earlier measurements in other buffer systems (Oza et al., 2003; Torrie et al., 2009). Major alterations were a threefold higher K_i value for GSH, a rise of the K_m value for Spd from 139 to 687 μM and a reduction of the k_{cat} from 5.2 to 2.8 s^{-1}. These

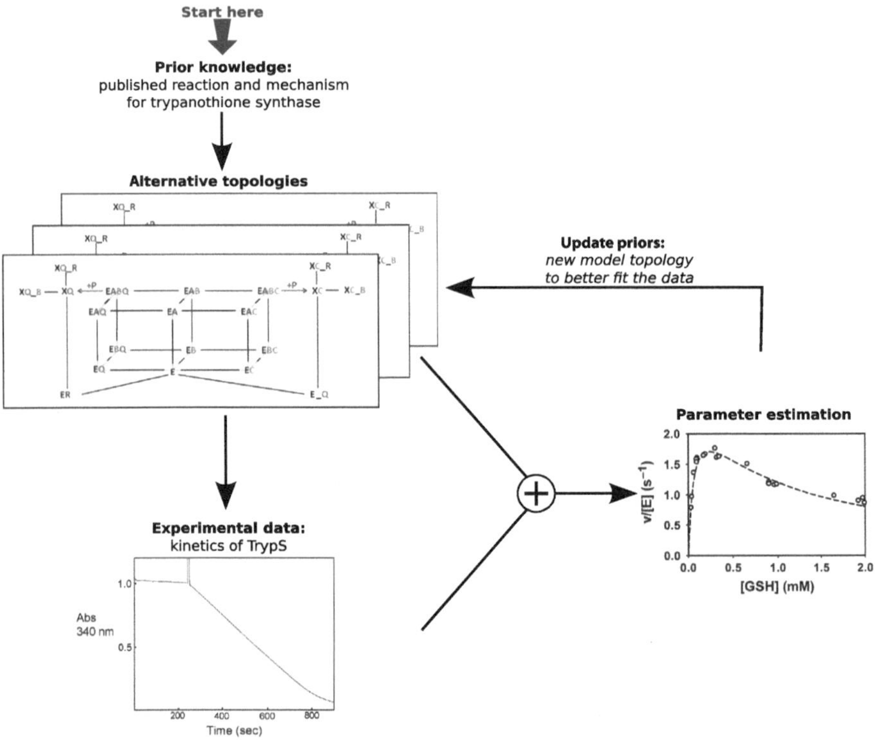

Figure 3.8 Workflow applied for dissecting the kinetic mechanism of trypanothione synthetase. *Modified with permission from Leroux et al. (2013). Copyright 2013 American Society for Biochemistry and Molecular Biology.*

observations are further demonstration of the importance of mimicking the physiological environment.

4. TRANSCRIPTION/TRANSLATION EXTENSIONS OF THE MODEL

One of the aims of the SilicoTryp project is to include transcription and translation in the model, resulting in dynamic enzyme levels, instead of using fixed V_{max} values for all enzymes. This will be done using a module of four differential equations for each enzyme (see Fig. 3.9), based on the previously published model of phosphoglycerol kinase transcription and translation (Haanstra, Stewart, et al., 2008). Regulation of enzyme concentrations in *T. brucei* is typically achieved by controlling the pre-mRNA-processing rate ($k_{splicing}$ in Fig. 3.9), the translation rate ($k_{translation}$) and/or

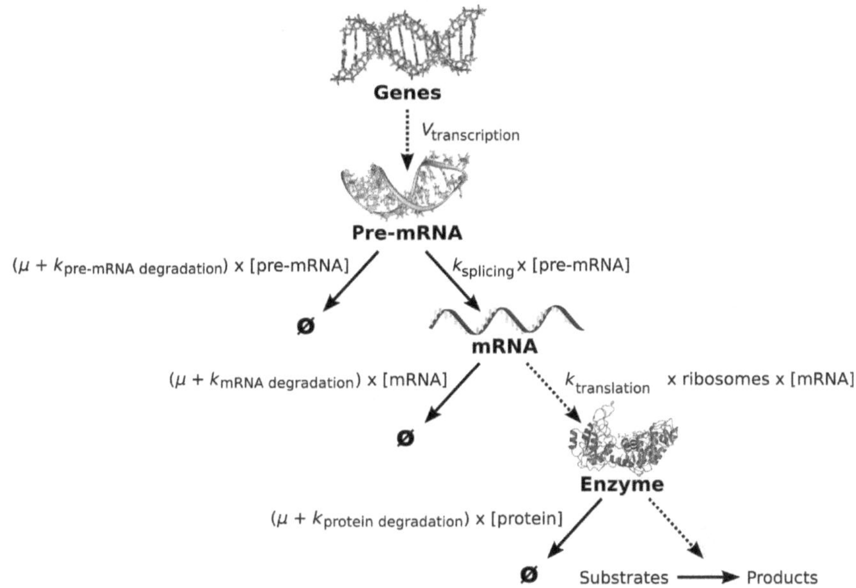

Figure 3.9 Transcription/translation module that will be added for every enzyme of the model. The module is based on the model published by Haanstra, Stewart, et al. (2008). The rate constants of the various processes involved are indicated by k, and μ are the specific growth rate of the trypanosomes, and *ribosomes* represents the number of ribosomes per molecule of mRNA. (See the color plate.)

the degradation rates. To measure the pre-mRNA and mRNA degradation and processing rates, RNAseq experiments were done and are currently being analysed.

4.1. The pre-mRNA-processing rates

Genes in trypanosomes are transcribed as polycistronic precursor messages (pre-mRNA) that are destined for either trans-splicing into mature mRNA or degradation. In order to measure the rate of pre-mRNA-processing for each gene, cells were treated with Actinomycin D for 5 min to inhibit mRNA transcription. RNA was then extracted and, after being depleted from rRNA, two biological replicates and two controls were subjected to transcriptome-wide deep sequencing (RNAseq, an average of 50 million reads of length 50mer, using Illumina HiSeq 2000). A decrease in pre-mRNA read counts in treated cells will be due to the splicing of the pre-mRNA in the absence of transcription. The ratio of pre-mRNA read counts (treated cells over untreated) was then used to estimate the rate of mRNA processing. A difficulty in differentiating between reads from pre-mRNA sequences and those due to spurious transcription events in the same region, led us to limit our analysis to the 20 nucleotides immediately upstream of known splice acceptor sites. Although this had the drawback of limiting the number of examined regions, this filter increases our confidence about the origin of the reads.

Preliminary results indicate that 50% of the transcripts have a half-life of less than 2 min, but some showed much longer apparent half-lives. We speculate that this is due to persistence of bi-cistronic RNAs, perhaps in the cytoplasm (Fadda, A. et al., in preparation).

4.2. mRNA decay rates

In order to measure mRNA degradation rates, cells were treated with both Actinomycin D and Sinefungin to inhibit both transcription and splicing. RNA was extracted at several time points after transcription inhibition, depleted from rRNA and subjected to deep sequencing (an average of 25 million 75mer reads, or 200 million 50mer reads, using Illumina GAIIX and HiSeq 2000, respectively). Since all trypanosome mRNAs carry the same 39-nt spliced leader (SL) at the 5′-end, the abundance of the SL can serve as a proxy for the abundance of total mRNA. This allowed us to normalise the read counts using the total mRNA decay pattern across the same time points. Then, amounts of total mRNA were measured relative to the steady state (Fadda, Färber, Droll, & Clayton, 2013), and mRNA half-lives were first estimated using an exponential model.

It has been shown, however, that in a transcription inhibition experiment, while the majority of transcripts decay exponentially, others exhibit a slow-to-fast or fast-to-slow decay pattern. In our work, we see the same pattern. Therefore, we used a Markov chain model developed earlier by Deneke, Lipowsky, and Valleriani (2013) to analyse the data for transcripts that do not show exponential kinetics. Transcripts exhibiting non-exponential decay are presumed to be degraded in a multi-step reaction, in which more than one rate-limiting step exists, and where the age of the transcript becomes relevant to its rate of degradation. This is now further studied with a particular focus on the mRNAs encoding enzymes of the glycolytic and redox pathways, many of which show non-exponential decay.

The mRNA kinetic data are furthermore being complemented with information on the translation rates for each transcript, using a ribosome profiling (Ingolia, Ghaemmaghami, Newman, & Weissman, 2009) strategy.

The number of ribosomes on each mRNA is determined by treating cell lysates with RNase to digest the mRNA except those stretches protected by ribosomes. The protected RNA fragments are then purified and sequenced, and the rate of translation estimated by assuming a linear relationship between the ribosome number on a transcript and the rate of translation.

5. EXPERIMENTAL DATA TO REDUCE MODEL UNCERTAINTY AND REFINE PREDICTIONS

In order to reduce our uncertainty about both model topology and parameter values, further experimental data are needed, in addition to the transcription dynamics work described above. Data collection within SilicoTryp focused initially on steady-state metabolite concentrations and fluxes, which are directly informative for model refinement.

In addition, a collection of RNAi mutants was constructed with the aim of providing additional information on the system's behaviour upon perturbation.

5.1. Measurements of concentrations and fluxes

Intracellular and extracellular concentrations and fluxes were measured using LC–MS. However, mass spectrometry intensities are not always linearly proportional to the concentrations due to ion suppression effects (Annesley, 2003). In addition, variations in instrumental response and degradation of metabolites of interest during sample preparation can also result in a biased quantitative result (Vuckovic, 2012). To overcome this problem, we need to spike a fixed amount of the ^{13}C-labelled metabolite into each calibration sample (Mashego et al., 2004). Being chemically identical but having a different mass, this compound will be similarly ion suppressed within the matrix entering the mass spectrometer, but distinguished from unlabelled compound since its mass is slightly different. Therefore, the ratios of intensities of the ^{12}C-metabolite over the ^{13}C-metabolite are directly proportional to the concentrations of ^{12}C-metabolite. Using a calibration curve, the concentration of the metabolite can be determined in any sample if the same amount of ^{13}C-metabolite that has been used for the calibration curve is added to the sample. However, ^{13}C-labelled compounds are expensive and difficult to obtain. Therefore, uniformly (U)-^{13}C-labelled *E. coli* metabolite extracts, obtained by growing *E. coli* in M9 minimal medium with (U)-^{13}C-labelled glucose as the sole carbon source (Kiefer, Portais, & Vorholt, 2008), were used as standards in the SilicoTryp project. This approach also offers the advantage that it potentially allows simultaneous quantification of all metabolites that can be detected in both *Escherichia coli* and *T. brucei*.

Using this method, we quantified 44 intracellular metabolites and 32 metabolites in the growth medium over a 57-h time course (starting from the low cell density of 3800 cells/ml to maintain the cells in exponential growth throughout the time course). Based on the 32 medium metabolite dynamics, we could quantify the fluxes of six metabolites that are constantly produced or consumed by the cells using the absolute concentrations and the cell densities at the same time points. Seven metabolites did not show any significant changes in concentration over time; for example, ornithine levels did not seem to change in the spent medium, which was unexpected since previous work had demonstrated that ornithine is taken up by the cells. Nineteen metabolites, all amino acids or nucleotides, showed changes of concentrations that are not compatible with constant production or consumption by the cells, but are compatible with the presence of peptidases and nucleosidases which degrade proteins and nucleic acids to their corresponding monomers. Indeed, it is known that *T. brucei* excretes peptidases and nucleosidases into the medium (Bossard, Cuny, & Geiger, 2013; Geiger et al., 2010; Knowles, Black, & Whitelaw, 1987).

From these results, and the ^{13}C-labelling experiments (see above), we could, for example, determine that about 2–3% of the pyruvate produced by cells grown in the simplified medium (Creek et al., 2013; see Section 5.2) is converted to alanine before being excreted. These results can now be used to extend the model and reduce our uncertainty about topology and parameter values.

5.2. Perturbations: RNAi cell lines

In order to provide information on the effect of reduced level of selected enzymes on the levels of mRNA, protein and the metabolome, we generated 28 stable RNA interference (RNAi) cell lines: 8 of them targeting genes of the polyamine pathway (the part of the trypanothione pathway that leads to the production of Spd in Fig. 3.7), 12 of the redox metabolism and 8 of the pentose phosphate pathway (see Table 3.2).

The RNAi cell lines were generated in monomorphic 2T1 *Trypanosoma brucei brucei*, selected among other available cell lines as it provides a unique target sequence at a single ribosomal RNA locus, and has been validated for robust inducible expression and has a higher RNAi efficiency (Alsford, Kawahara, Glover, & Horn, 2005). To confirm the specific down regulation

Table 3.2 List of RNAi mutants constructed and the previously published phenotypes

Gene ID	Protein	Growth reduction/ phenotype	References
Polyamine pathway			
Tb09.v1.0380	Spermidine synthase	Spermidine auxotrophy, depletion of $T(SH)_2$ and cell death	Xiao, McCloskey, and Phillips (2009)
Tb11.01.5300	Ornithine decarboxylase	Depletion of reduced thiols and cell growth arrest	Fairlamb, Henderson, Bacchi, and Cerami (1987)
Tb927.6.4410	*S*-adenosylmethionine decarboxylase	60–90%	Alsford et al. (2011)
Tb927.6.4470	Prozyme, activator of SAMDC	100%	Alsford et al. (2011)
Tb927.6.4840	Methionine adenosyltransferase	No effect	Alsford et al. (2011)
Tb927.6.4890	*S*-adenosylmethionine synthase	Unknown	Alsford et al. (2011)
Tb927.8.1910	Acetylornithine deacetylase	No effect	Alsford et al. (2011)
Pentose phosphate pathway and related enzymes			
Tb927.10.2490	Glucose-6-phosphate 1-dehydrogenase	Growth arrest, cell death	Cordeiro, Thiemann, and Michels (2009)
Tb11.02.4200	6-Phosphogluconolactonase	None	Alsford et al. (2011)
Tb09.211.3180	6-Phosphogluconate dehydrogenase	70–100%	Alsford et al., (2011)[a]
Tb927.8.6170	Transketolase	None	Alsford et al. (2011)
Tb927.10.12210	Ribulose-5-phosphate 3-epimerase	None	Alsford et al. (2011)
Tb11.03.0090	Ribokinase	70–80%	Alsford et al. (2011)

Continued

Table 3.2 List of RNAi mutants constructed and the previously published phenotypes—cont'd

Gene ID	Protein	Growth reduction/ phenotype	References
Tb927.4.1350	Glyoxalase II	20–80%	Alsford et al. (2011) and Wendler, Irsch, Rabbani, Thornalley, and Krauth-Siegel (2009)
Tb11.01.0700	Ribose 5-phosphate isomerase	>90%	Alsford et al. (2011)
Redox metabolism			
Tb927.3.3760	Tryparedoxin	Growth arrest and enhanced sensitivity against hydrogen peroxide	Alsford et al. (2011) and Comini, Krauth-Siegel, and Flohé (2007)
Tb09.160.2020	Thioredoxin	None	Alsford et al. (2011) and Schmidt, Clayton, and Krauth-Siegel (2002)
Tb927.2.4370	Trypanothione synthetase	Depletion of Gsp and $T(SH)_2$, growth arrest, loss of viability, enhanced sensitivity against hydroperoxides and drugs	Comini et al. (2004) and Ariyanayagam, Oza, Guther, and Fairlamb (2005)
Tb927.7.4000	Glutathione synthetase	10–80%	Alsford et al. (2011)
Tb927.10.12370	γ-Glutamylcysteine synthetase	Depletion of GSH and $T(SH)_2$, cell death and loss of virulence	Huynh, Huynh, Harmon, and Phillips (2003)

Table 3.2 List of RNAi mutants constructed and the previously published phenotypes—cont'd

Gene ID	Protein	Growth reduction/ phenotype	References
Tb927.10.10390	Trypanothione reductase	Growth arrest, loss of viability and virulence	Krieger et al. (2000)
Tb11.01.7560	Glutathione peroxidase, putative	>90%	Alsford et al. (2011)
Tb11.12.0016	Glutathionylspermidine synthetase	No effect	Alsford et al. (2011)
Tb09.160.4250	Peroxiredoxin-type tryparedoxin peroxidase (cytosolic)	Loss of viability and enhanced sensitivity against hydrogen peroxide	Alsford et al. (2011) and Wilkinson, Horn, Prathalingam, and Kelly (2003)
Tb927.4.2450	Protein disulfide isomerase I	15–60%	Alsford et al. (2011)
Tb927.3.4240	Thioredoxin, putative	40–80%	Alsford et al. (2011)
Tb927.5.950	Monothiol glutaredoxin 3	50–100%	Alsford et al. (2011)
Tb927.8.1990	Peroxiredoxin-type tryparedoxin peroxidase (mitochondrial)	None	Alsford et al., (2011) and Wilkinson et al. (2003)

[a]Kerkhoven et al. (2013).

of these enzymes at the protein level, RNAi cell lines where antibodies are available were selected, induced and tested by Western blot. So far, all RNAi lines tested by Western blot show down regulation of the protein of interest. The effects of ablating some of these enzymes have already been published, but most remain to be fully characterised (see Table 3.2). From the pentose phosphate pathway, only the glucose 6-phosphate 1-dehydrogenase (Cordeiro et al., 2009) and the 6-phosphogluconate dehydrogenase (Hanau, Rippa, Bertelli, Dallocchio, & Barrett, 1996) have been characterised previously.

Most of the evaluation of phenotypes in previous work has been carried out in a medium containing a vast excess of many metabolites. Comparing fresh and spent medium metabolomes revealed that most metabolites, however, are not consumed by trypanosomes. We therefore created a new medium from which those metabolites not consumed by trypanosomes were removed (Creek et al., 2013). Parasites grew equally well in this medium, which better reflects serum concentrations of many metabolites. Remarkably, experiments in the simplified medium revealed phenotypic effects that were hidden in the classical medium. For example, *T. brucei* bloodstream forms became over 100-fold more sensitive to antifolate drugs in the absence of the high concentrations of folate that are present in the classical medium (Creek et al., 2013).

The individual stable RNAi cell lines available now for each of our main genes of interest will allow us to confirm these results and also measure the changes in metabolites upon depletion of any particular enzyme as well as the effects, if any, on mRNA or protein levels of other enzymes. These perturbation responses will then inform the model debugging and refinement process.

6. CONCLUSION

The SilicoTryp project aimed at extending the core glycolysis model of the unicellular parasite *T. brucei* with the trypanothione pathway, and the translation and transcription of the enzymes involved. All of this work required newly developed modelling methods that explicitly take the multiple levels of biological and uncertainty into account. Hence, we first documented in detail the sources of the parameters of the existing core glycolysis model. This detailed information allowed us to re-evaluate this model taking our uncertainty about the parameter values into account (Achcar et al., 2012). In addition, we also questioned the topology of this model by analysing and comparing alternative possibilities (Achcar et al., 2013). These analyses formalised what our current levels of uncertainty are regarding the core glycolysis model. In the future, we aim at reducing these uncertainties using Bayesian inference methods, which use new experimental data to update our belief on parameter values and model topology (see Fig. 3.1). Because of the complexity of the models we are dealing with, we developed a new state-of-the-art variance reduction method specifically for this purpose (Papamarkou et al., 2014).

Although modelling attempts have been made for part of this pathway both in *T. brucei* (Gu, Reid, Higham, & Gilbert, 2013) and *T. cruzi* (Olin-Sandoval et al., 2012), working towards the introduction of the trypanothione pathway introduced a new set of challenges. First, introducing the pentose phosphate pathway, the link between glycolysis and the trypanothione pathway, revealed new uncertainties about the topology of this pathway in *T. brucei*. We were nevertheless able to predict relevant behaviours of the system based on the alternative versions of the model (Kerkhoven et al., 2013). Then, to be able to introduce the trypanothione pathway *per se*, we again had to clarify some topological questions. We were able to do so using ^{13}C-labelled metabolites and LC–MS. For example, we showed that MTA is not recycled to methionine in *T. brucei*. However, deciphering the origin of ornithine *in vivo* will require further investigation. In addition to investigating the topology of this pathway, new kinetic parameters of the enzymes involved have to be measured. To reduce our uncertainty about this new kinetics, we designed a new *in vivo*-like buffer. Using this new assay system and a method based on a strong interaction between modellers and experimentalists, we deciphered in detail the complex mechanism of the trypanothione synthetase, the last enzyme of the trypanothione biosynthesis pathway (Leroux et al., 2013).

Introducing the possibility of dynamic regulation to these pathways at the transcriptional and translational levels requires new sets of parameters that have to be measured using specific methodologies. These regulatory steps will be introduced to the models using a module of four differential equations based on the previously published model of PGK transcription and translation (Haanstra, Stewart, et al., 2008). We designed appropriate methodology to measure the parameters needed based on deep sequencing (RNAseq).

All of these new additions introduced many new parameters and additional topological uncertainties. To enable reduction of these uncertainties using a Bayesian methodology, we collected additional data about metabolite concentrations and fluxes, using a new method based on LC–MS data. Intracellular and extracellular concentrations and fluxes were collected in the wild type cell line. In the future, we also plan to collect metabolite concentrations in mutant cell lines to produce new information about the system under perturbation. Therefore, 28 RNAi cell lines targeting critical enzymes in our system were constructed and verified. We will now be able to grow these cell lines in uniform controlled conditions to characterise the metabolomic and transcriptomic response to these perturbations.

This work opens many novel perspectives: our main aim now is to apply our newly developed Bayesian inference methods to our models and data, starting with small models and gradually increasing model size. While the first small models will be analysed, new enzymes kinetics will be acquired, but also the deep sequencing data will be analysed and the RNAi mutants studied. Each additional dataset acquired will allow us to reduce our uncertainty on some part of the model. We argue that the conceptual framework we are developing here to study the metabolism of *T. brucei*, including the explicit quantitative consideration of topological and parameter uncertainty, can be applied to study any biological system and offers powerful new approaches to enable the robust and predictive simulation of cellular function.

ACKNOWLEDGEMENTS

This work was supported by the Scottish Universities Life Sciences Alliance (SULSA), the Federal Ministry of Education and Research (Germany), the Netherlands Organisation for Scientific Research (NWO, The Netherlands) and the Biotechnology and Biological Sciences Research Council (BBSRC, United Kingdom), within the framework of the SysMO2-funded SilicoTryp project. BMB received a Rosalind Franklin Fellowship from the University of Groningen.

REFERENCES

Achcar, F., Barrett, M. P., & Breitling, R. (2013). Explicit consideration of topological and parameter uncertainty gives new insights into a well-established model of glycolysis. *The FEBS Journal, 280*, 4640–4651.

Achcar, F., Kerkhoven, E. J., Bakker, B. M., Barrett, M. P., & Breitling, R. The SilicoTryp Consortium. (2012). Dynamic modelling under uncertainty: The case of *Trypanosoma brucei* energy metabolism. *PLoS Computational Biology, 8*(1), e1002352.

Albert, M. A., Haanstra, J. R., Hannaert, V., Van Roy, J., Opperdoes, F. R., Bakker, B. M., et al. (2005). Experimental and in silico analyses of glycolytic flux control in bloodstream form *Trypanosoma brucei*. *Journal of Biological Chemistry, 280*(31), 28306–28315.

Alsford, S., Kawahara, T., Glover, L., & Horn, D. (2005). Tagging a T. brucei RRNA locus improves stable transfection efficiency and circumvents inducible expression position effects. *Molecular and Biochemical Parasitology, 144*(2), 142–148.

Alsford, S., Turner, D. J., Obado, S. O., Sanchez-Flores, A., Glover, L., Berriman, M., et al. (2011). High-throughput phenotyping using parallel sequencing of RNA interference targets in the African trypanosome. *Genome Research, 21*(6), 915–924.

Annesley, T. M. (2003). Ion suppression in mass spectrometry. *Clinical Chemistry, 49*(7), 1041–1044.

Ariyanayagam, M. R., Oza, S. L., Guther, M. L. S., & Fairlamb, A. H. (2005). Phenotypic analysis of trypanothione synthetase knockdown in the African trypanosome. *The Biochemical Journal, 391*(Pt 2), 425–432.

Bakker, B. M., Mensonides, F. I., Teusink, B., van Hoek, P., Michels, P. A. M., & Westerhoff, H. V. (2000). Compartmentation protects trypanosomes from the dangerous design of glycolysis. *Proceedings of the National Academy of Sciences of the United States of America, 97*(5), 2087–2092.

Bakker, B. M., Michels, P. A. M., Opperdoes, F. R., & Westerhoff, H. V. (1997). Glycolysis in bloodstream form *Trypanosoma brucei* can be understood in terms of the kinetics of the glycolytic enzymes. *Journal of Biological Chemistry, 272*(6), 3207–3215.

Bakker, B. M., Michels, P. A. M., Opperdoes, F. R., & Westerhoff, H. V. (1999). What controls glycolysis in bloodstream form *Trypanosoma brucei*? *Journal of Biological Chemistry, 274*(21), 14551–14559.

Bakker, B. M., Walsh, M. C., ter Kuile, B. H., Mensonides, F. I. C., Michels, P. A. M., Opperdoes, F. R., et al. (1999). Contribution of glucose transport to the control of the glycolytic flux in *Trypanosoma brucei*. *Proceedings of the National Academy of Sciences of the United States of America, 96*(18), 10098–10103.

Barrett, M. P., Bakker, B. M., & Breitling, R. (2010). Metabolomic systems biology of trypanosomes. *Parasitology, 137*(9), 1285–1290.

Barrett, M. P., Burchmore, R. J. S., Stich, A., Lazzari, J. O., Frasch, A. C., Cazzulo, J. J., et al. (2003). The trypanosomiases. *Lancet, 362*(9394), 1469–1480.

Bossard, G., Cuny, G., & Geiger, A. (2013). Secreted proteases of *Trypanosoma brucei gambiense*: Possible targets for sleeping sickness control? *Biofactors, 39*, 407–414.

Breitling, R., Achcar, F., & Takano, E. (2013). Modeling challenges in the synthetic biology of secondary metabolism. *ACS Synthetic Biology, 2*(7), 373–378.

Comini, M. A., Guerrero, S. A., Haile, S., Menge, U., Lünsdorf, H., & Flohé, L. (2004). Validation of *Trypanosoma brucei* trypanothione synthetase as drug target. *Free Radical Biology and Medicine, 36*(10), 1289–1302.

Comini, M. A., Krauth-Siegel, R. L., & Flohé, L. (2007). Depletion of the thioredoxin homologue tryparedoxin impairs antioxidative defence in African trypanosomes. *The Biochemical Journal, 402*(1), 43–49.

Cordeiro, A. T., Thiemann, O. H., & Michels, P. A. M. (2009). Inhibition of *Trypanosoma brucei* glucose-6-phosphate dehydrogenase by human steroids and their effects on the viability of cultured parasites. *Bioorganic and Medicinal Chemistry, 17*(6), 2483–2489.

Creek, D. J., Nijagal, B., Kim, D. H., Rojas, F., Matthews, K. R., & Barrett, M. P. (2013). Metabolomics guides rational development of a simplified cell culture medium for drug screening against *Trypanosoma brucei*. *Antimicrobial Agents and Chemotherapy, 57*(6), 2768–2779.

Deneke, C., Lipowsky, R., & Valleriani, A. (2013). Complex degradation processes lead to non-exponential decay patterns and age-dependent decay rates of messenger RNA. *PLoS One, 8*(2), e55442.

Fadda, A., Färber, V., Droll, D., & Clayton, C. (2013). The roles of 3′-exoribonucleases and the exosome in trypanosome mRNA degradation. *RNA, 19*(7), 937–947.

Fairlamb, A. H., Henderson, G. B., Bacchi, C. J., & Cerami, A. (1987). In vivo effects of difluoromethylornithine on trypanothione and polyamine levels in bloodstream forms of *Trypanosoma brucei*. *Molecular and Biochemical Parasitology, 24*(2), 185–191.

Fuad, F. A. A., Fothergill-Gilmore, L. A., Nowicki, M. W., Eades, L. J., Morgan, H. P., McNae, I. W., et al. (2011). Phosphoglycerate mutase from *Trypanosoma brucei* is hyperactivated by cobalt *in vitro*, but not *in vivo*. *Metallomics, 3*(12), 1310–1317.

Geiger, A., Hirtz, C., Bécue, T., Bellard, E., Centeno, D., Gargani, D., et al. (2010). Exocytosis and protein secretion in Trypanosoma. *BMC Microbiology, 10*, 20.

Gu, X., Reid, D., Higham, D. J., & Gilbert, D. (2013). Mathematical modelling of polyamine metabolism in bloodstream-form *Trypanosoma brucei*: An application to drug target identification. *PLoS One, 8*(1), e53734.

Gualdron-López, M., Vapola, M. H., Miinalainen, I. J., Hiltunen, J. K., Michels, P. A. M., & Antonenkov, V. D. (2012). Channel-forming activities in the glycosomal fraction from the bloodstream form of *Trypanosoma brucei*. *PLoS One*, 7(4), e34530.

Haanstra, J. R., Stewart, M., Luu, V. D., van Tuijl, A., Westerhoff, H. V., Clayton, C., et al. (2008). Control and regulation of gene expression: Quantitative analysis of the expression of phosphoglycerate kinase in bloodstream form *Trypanosoma brucei*. *Journal of Biological Chemistry*, *283*(5), 2495–2507.

Haanstra, J. R., van Tuijl, A., Kessler, P., Reijnders, W., Michels, P. A. M., Westerhoff, H. V., et al. (2008). Compartmentation prevents a lethal turbo-explosion of glycolysis in trypanosomes. *Proceedings of the National Academy of Sciences of the United States of America*, *105*(46), 17718–17723.

Hanau, S., Rippa, M., Bertelli, M., Dallocchio, F., & Barrett, M. P. (1996). 6-Phosphogluconate dehydrogenase from *Trypanosoma brucei*. Kinetic analysis and inhibition by trypanocidal drugs. *European Journal of Biochemistry*, *240*(3), 592–599.

Häusler, T., Stierhof, Y. D., Wirtz, E., & Clayton, C. (1996). Import of a DHFR hybrid protein into glycosomes in vivo is not inhibited by the folate-analogue aminopterin. *Journal of Cell Biology*, *132*(3), 311–324.

Helfert, S., Estévez, A. M., Bakker, B., Michels, P., & Clayton, C. (2001). Roles of triosephosphate isomerase and aerobic metabolism in *Trypanosoma brucei*. *The Biochemical Journal*, *357*(Pt 1), 117–125.

Huynh, T. T., Huynh, V. T., Harmon, M. A., & Phillips, M. A. (2003). Gene knockdown of gamma-glutamylcysteine synthetase by RNAi in the parasitic protozoa *Trypanosoma brucei* demonstrates that it is an essential enzyme. *Journal of Biological Chemistry*, *278*(41), 39794–39800.

Ingolia, N. T., Ghaemmaghami, S., Newman, J. R. S., & Weissman, J. S. (2009). Genome-wide analysis in vivo of translation with nucleotide resolution using ribosome profiling. *Science*, *324*(5924), 218–223.

Kerkhoven, E. J., Achcar, F., Alibu, V. P., Burchmore, R. J., Gilbert, I. H., Trybiło, M., et al. (2013). Handling uncertainty in dynamic models: the pentose phosphate pathway in *Trypanosoma brucei*. *PLoS Computational Biolog*, *9*(12), e1003371.

Kiefer, P., Portais, J. C., & Vorholt, J. A. (2008). Quantitative metabolome analysis using liquid chromatography-high-resolution mass spectrometry. *Analytical Biochemistry*, *382*(2), 94–100.

Knowles, G., Black, S. J., & Whitelaw, D. D. (1987). Peptidase in the plasma of mice infected with *Trypanosoma brucei brucei*. *Parasitology*, *95*(Pt 2), 291–300.

Krieger, S., Schwarz, W., Ariyanayagam, M. R., Fairlamb, A. H., Krauth-Siegel, R. L., & Clayton, C. (2000). Trypanosomes lacking trypanothione reductase are avirulent and show increased sensitivity to oxidative stress. *Molecular Microbiology*, *35*(3), 542–552.

Leroux, A. E., Haanstra, J. R., Bakker, B. M., & Krauth-Siegel, R. L. (2013). Dissecting the catalytic mechanism of *Trypanosoma brucei* trypanothione synthetase by kinetic analysis and computational modelling. *The Journal of Biological Chemistry*, *288*, 23751–23764.

Martens-Lobenhoffer, J., Postel, S., Tröger, U., & Bode-Böger, S. M. (2007). Determination of ornithine in human plasma by hydrophilic interaction chromatography–tandem mass spectrometry. *Journal of Chromatography B: Analytical Technologies in the Biomedical and Life Sciences*, *855*(2), 271–275.

Mashego, M. R., Wu, L., Van Dam, J. C., Ras, C., Vinke, J. L., Van Winden, W. A., et al. (2004). MIRACLE: Mass isotopomer ratio analysis of u-13C-labeled extracts. A new method for accurate quantification of changes in concentrations of intracellular metabolites. *Biotechnology and Bioengineering*, *85*(6), 620–628.

Michels, R. A. M., Moyersoen, J., Krazy, H., Galland, N., Herman, M., & Hannaert, V. (2005). Peroxisomes, glyoxysomes and glycosomes (review). *Molecular Membrane Biology*, *22*(1–2), 133–145.

Moreno, B., Urbina, J. A., Oldfield, E., Bailey, B. N., Rodrigues, C. O., & Docampo, R. (2000). ^{31}P NMR spectroscopy of *Trypanosoma brucei*, *Trypanosoma cruzi*, and *Leishmania major*: Evidence for high levels of condensed inorganic phosphates. *Journal of Biological Chemistry*, *275*(37), 28356–28362.

Nolan, D. P., & Voorheis, H. P. (2000a). Hydrogen ion gradients across the mitochondrial, endosomal and plasma membranes in bloodstream forms of *Trypanosoma brucei* solving the three-compartment problem. *European Journal of Biochemistry*, *267*(15), 4601–4614.

Nolan, D. P., & Voorheis, H. P. (2000b). Factors that determine the plasma-membrane potential in bloodstream forms of *Trypanosoma brucei*. *European Journal of Biochemistry*, *267*(15), 4615–4623.

Olin-Sandoval, V., González-Chávez, Z., Berzunza-Cruz, M., Martínez, I., Jasso-Chávez, R., Becker, I., et al. (2012). Drug target validation of the trypanothione pathway enzymes through metabolic modelling. *The FEBS Journal*, *279*(10), 1811–1833.

Oza, S. L., Ariyanayagam, M. R., Aitcheson, N., & Fairlamb, A. H. (2003). Properties of trypanothione synthetase from *Trypanosoma brucei*. *Molecular and Biochemical Parasitology*, *131*(1), 25–33.

Papamarkou, T., Mira, A., & Girolami, M. (2014). Zero variance differential geometric Markov Chain Monte Carlo algorithms. *Bayesian Analysis*, *9*(1), 97–128.

Schmidt, A., Clayton, C. E., & Krauth-Siegel, R. L. (2002). Silencing of the thioredoxin gene in Trypanosoma brucei brucei. *Molecular and Biochemical Parasitology*, *125*(1–2), 207–210.

Siegel, T. N., Gunasekera, K., Cross, G. A. M., & Ochsenreiter, T. (2011). Gene expression in *Trypanosoma brucei*: Lessons from high-throughput RNA sequencing. *Trends in Parasitology*, *27*(10), 434–441.

Thissen, J. A., & Wang, C. C. (1991). Maintenance of internal pH and an electrochemical gradient in *Trypanosoma brucei*. *Experimental Parasitology*, *72*(3), 243–251.

Torrie, L. S., Wyllie, S., Spinks, D., Oza, S. L., Thompson, S., Harrison, J. R., et al. (2009). Chemical validation of trypanothione synthetase: A potential drug target for human trypanosomiasis. *Journal of Biological Chemistry*, *284*(52), 36137–36145.

van Eunen, K., Bouwman, J., Daran-Lapujade, P., Postmus, J., Canelas, A. B., Mensonides, F. I. C., et al. (2010). Measuring enzyme activities under standardized in vivo-like conditions for systems biology. *The FEBS Journal*, *277*(3), 749–760.

van Eunen, K., Kiewiet, J. A. L., Westerhoff, H. V., & Bakker, B. M. (2012). Testing biochemistry revisited: How in vivo metabolism can be understood from in vitro enzyme kinetics. *PLoS Computational Biology*, *8*(4), e1002483.

Vincendeau, P., Gobert, A. P., Daulouède, S., Moynet, D., & Djavad Mossalayi, M. (2003). Arginases in parasitic diseases. *Trends in Parasitology*, *19*(1), 9–12.

Vincent, I. M., Creek, D. J., Burgess, K., Woods, D. J., Burchmore, R. J. S., & Barrett, M. P. (2012). Untargeted metabolomics reveals a lack of synergy between nifurtimox and eflornithine against Trypanosoma brucei. *PLoS Neglected Tropical Diseases*, *6*(5), e1618.

Vuckovic, D. (2012). Current trends and challenges in sample preparation for global metabolomics using liquid chromatography-mass spectrometry. *Analytical and Bioanalytical Chemistry*, *403*(6), 1523–1548.

Vyshemirsky, V., & Girolami, M. A. (2008). Bayesian ranking of biochemical system models. *Bioinformatics*, *24*(6), 833–839.

Wendler, A., Irsch, T., Rabbani, N., Thornalley, P. J., & Krauth-Siegel, R. L. (2009). Glyoxalase II does not support methylglyoxal detoxification but serves as a general trypanothione thioesterase in African trypanosomes. *Molecular and Biochemical Parasitology*, *163*(1), 19–27.

Wilkinson, S. R., Horn, D., Prathalingam, S. R., & Kelly, J. M. (2003). RNA interference identifies two hydroperoxide metabolizing enzymes that are essential to the bloodstream form of the African trypanosome. *Journal of Biological Chemistry*, *278*(34), 31640–31646.

Xiao, Y., McCloskey, D. E., & Phillips, M. A. (2009). RNA interference-mediated silencing of ornithine decarboxylase and spermidine synthase genes in *Trypanosoma brucei* provides insight into regulation of polyamine biosynthesis. *Eukaryotic Cell*, *8*(5), 747–755.

Xu, T. R., Vyshemirsky, V., Gormand, A., von Kriegsheim, A., Girolami, M., Baillie, G. S., et al. (2010). Inferring signaling pathway topologies from multiple perturbation measurements of specific biochemical species. *Science Signaling*, *3*(113), ra20.

AUTHOR INDEX

Note: Page numbers followed by "*f*" indicate figures and "*t*" indicate tables and "*np*" indicate footnotes.

A

B

C

D

J

K

L

W

X

Y

Z

SUBJECT INDEX

Note: Page numbers followed by "*f*" indicate figures and "*t*" indicate tables.

F

G

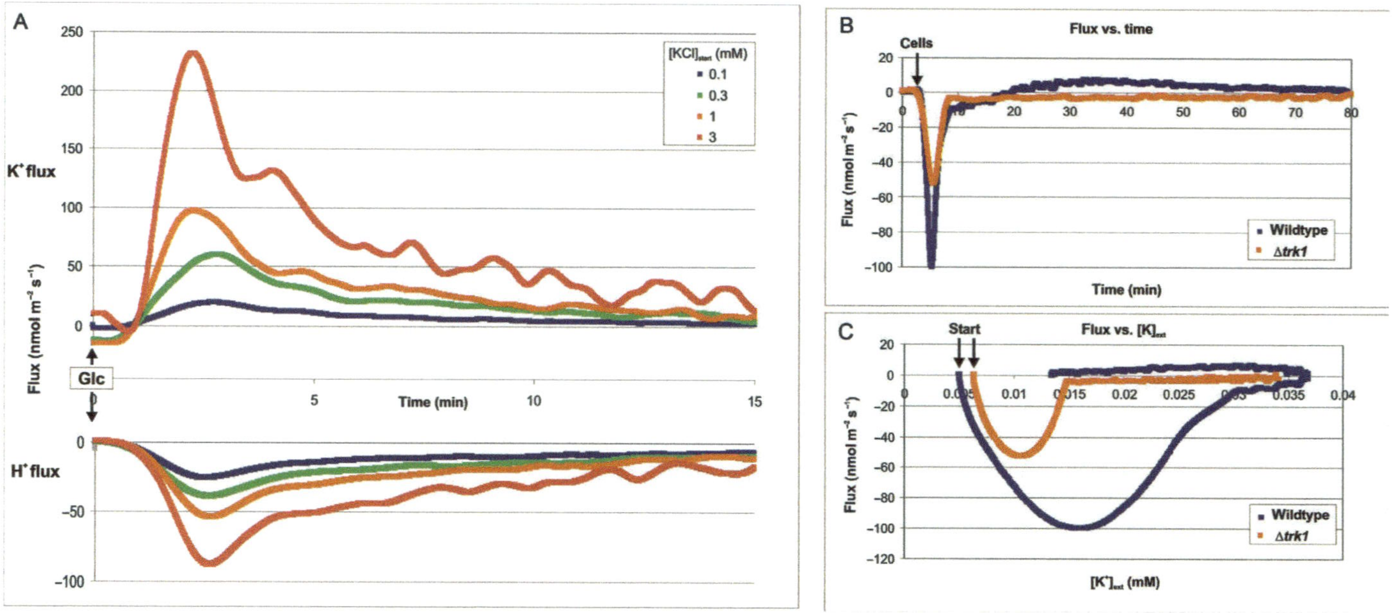

Joaquín Ariño *et al.*, Figure 1.6 Ion fluxes determined using FLISE. (A) Simultaneous recorded K^+ and H^+ fluxes at different added $[KCl]_{start}$ as indicated across the plasma membrane of BY4741 wild-type cells after starvation. Cells were energized by addition of glucose at time 0. Not only potassium uptake (positive fluxes) but also H^+ extrusion (negative fluxes) were dependent on $[K^+]_{ext}$. However, the fluxes do not sum up to 0. Whereas at low $[K]_{ext}$ proton efflux was larger than K^+ uptake, the ratio reverted at higher external K^+ concentrations. The apparent "oscillations" observed with 3 mM $[K^+]_{ext}$ are artefacts caused by smoothing and numerical calculation of the derivative. (B) Time course of K^+ efflux during potassium starvation. Cells were added to nominally K^+-free medium at the time indicated and flux was monitored for 80 min. Wild-type cells showed a transient efflux that eventually reverted to K^+ re-uptake. *trk1*Δ cells exhibited less efflux that continued during the course of the experiment. (C) Plot of K^+ fluxes shown in *B* versus $[K^+]_{ext}$. The external K^+ concentration had only increased to ~37 μM at the onset of re-uptake by wt cells and decreased to less than 15 μM.

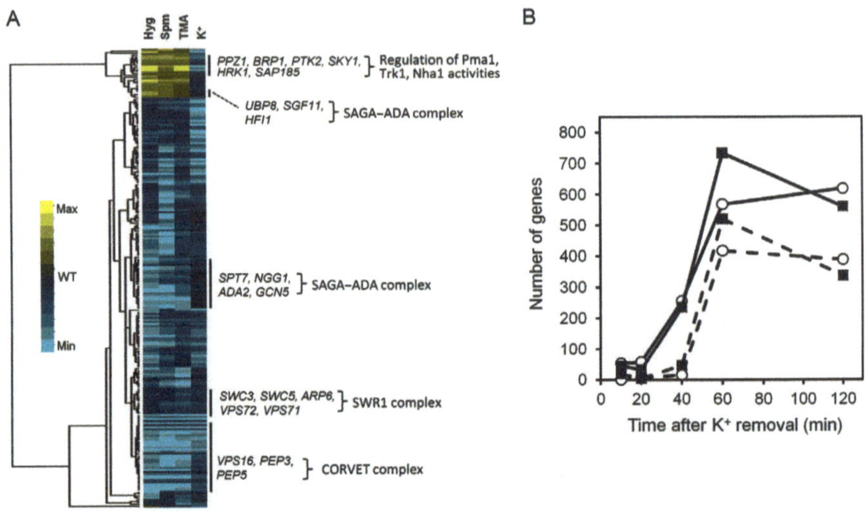

Joaquín Ariño *et al.*, Figure 1.7 The OMICS of potassium homeostasis. (A) Heat-map displaying clustering analysis (Cluster v3.0, settings, correlation uncentered and average linkage) of tolerance to drugs and potassium requirements for the 226 mutants selected in the screen reported by Barreto et al. (2011). Phenotypes similar to the wild-type (WT) strain are shown in black. Hyg, hygromycin B; Spm, spermine; TMA, tetramethylammonium; K^+, growth at limiting potassium (1 mM). See the above for details. (B) The short-term transcriptional response to potassium starvation. The number of genes induced (circles, continuous line) or repressed (squares, continuous line) at a given time after removal of potassium from the medium is plotted. Discontinuous lines denote the number of genes that shows the peak of induction or repression at a given time.

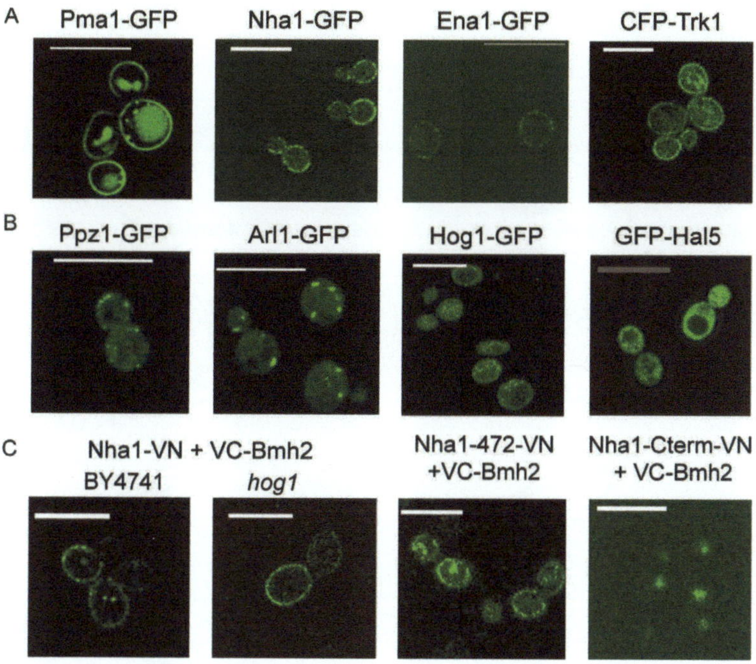

Joaquín Ariño *et al.*, Figure 1.8 Imaging ion transporters, their regulators and interactions. (A) Confocal laser scanning images of BY4741-expressing Pma1-GFP, Nha1-GFP, Ena1-GFP or CFP-Trk1. (B) Confocal laser scanning images of BY4741-expressing Ppz1-GFP, Arl1-GFP, Hog1-GFP or GFP-Hal5. (C) Confocal laser scanning images of BY4741-expressing Nha1-VN and VC-Bmh2, of *hog1*-expressing Nha1-VN and VC-Bmh2, of BY4741-expressing Nha1$_{1\text{-}472}$-VN and VC-Bmh2 and of BY4741-expressing Nha1$_{\text{C-term}}$-VN and VC-Bmh2 (for details, see Zahradka et al., 2012). Scale bar: 10 μm.

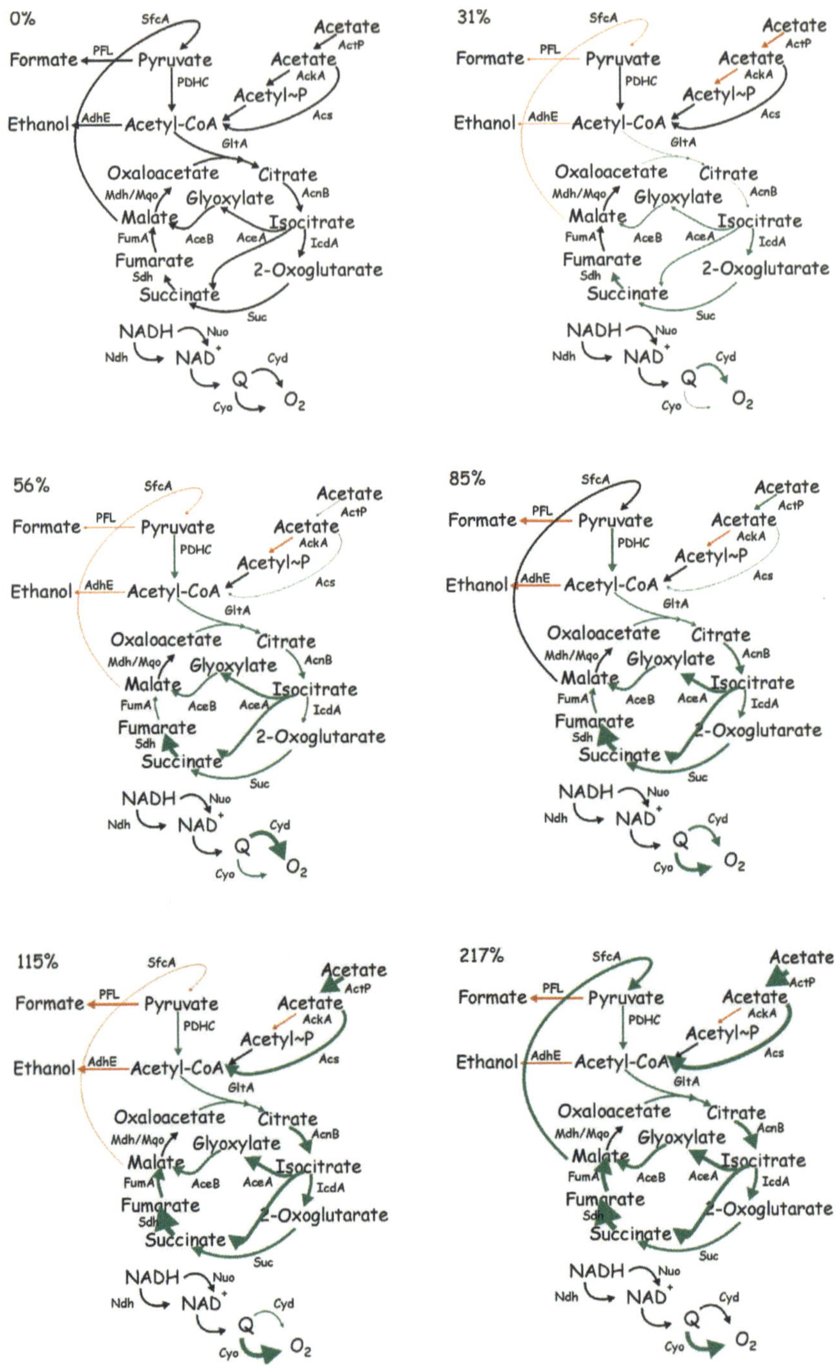

Katja Bettenbrock *et al.*, Figure 2.1 Changes in abundance of central metabolic transcripts in steady-state cultures grown at fixed pre-determined points on the aerobiosis scale. Transcripts with decreased abundance relative to the anaerobic steady state (0% aerobiosis) are shown in red; transcripts with increased abundance relative to the anaerobic steady state (0% aerobiosis) are shown in green. The width of each arrow provides an indication of the change in transcript abundance—the wider the arrow that greater the change. The point on the aerobiosis scale for each panel is indicated.

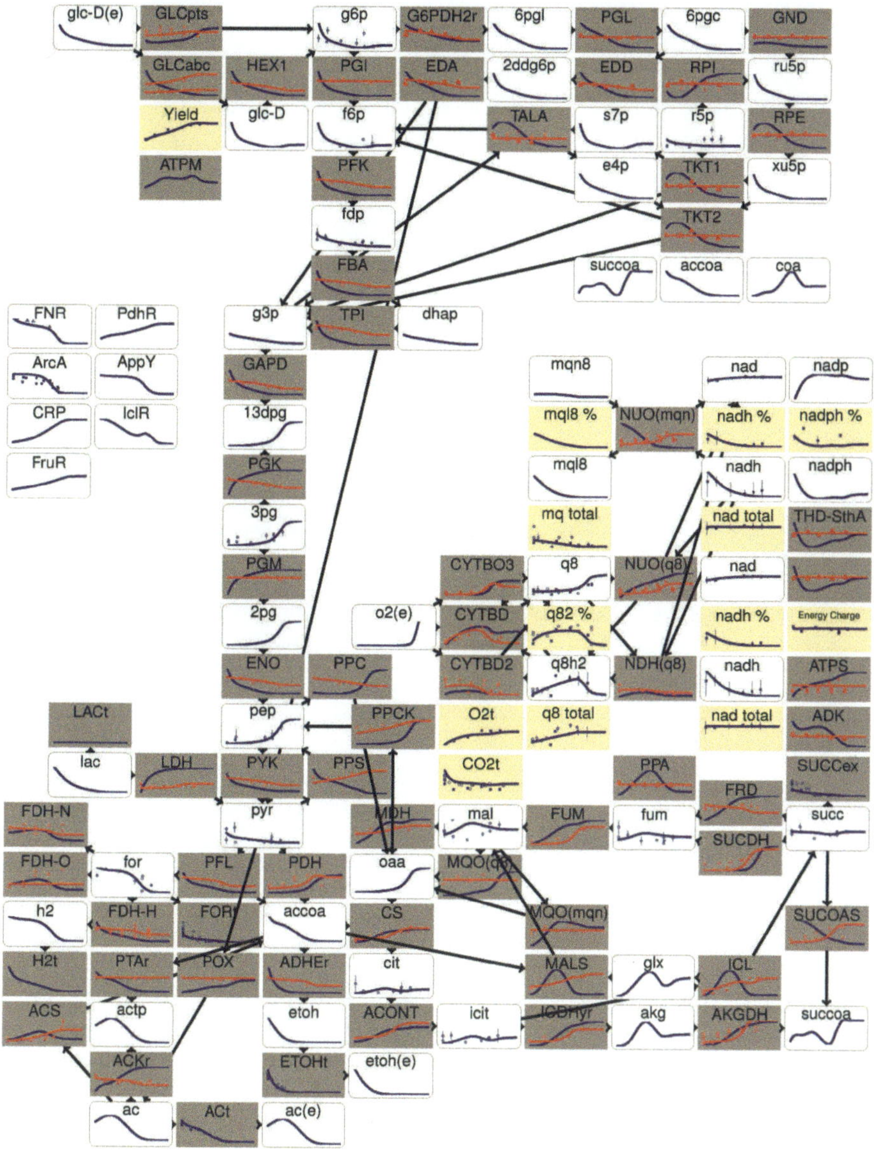

Katja Bettenbrock *et al.*, Figure 2.3 Simulation results of the central metabolism model. Shown are glucose uptake and glycolysis (upper left part of the figure), the fermentation pathways (lower left part), the citric acid cycle (lower right part), the electron transport chain (middle right part) and the pentose phosphate pathway (upper right part). White boxes represent metabolites and transcription factors. Blue lines and symbols in white boxes show simulations and measurements of concentrations, respectively. The labels of metabolites follow (Reed, Vo, Schilling, & Palsson, 2003): Grey boxes represent reactions. Blue lines and symbols in grey boxes show metabolic fluxes. Red lines and symbols in grey boxes show gene expression levels. The labels of the reactions follow the lables of (Reed et al., 2003). Yellow boxes show aggregated variables. The variables mql8%, q8h2%, nadh% and nadph% represent the degree of reduction of the menaquinone, ubiquinone, NADH and NADPH pools, respectively. The variables mq total, q8 total, nad total, nadp total are the total concentrations (reduced + oxidized) of the respective redox pairs. The variables O2t and CO2t are the biomass-specific exchange rates of oxygen and carbon dioxide, respectively.

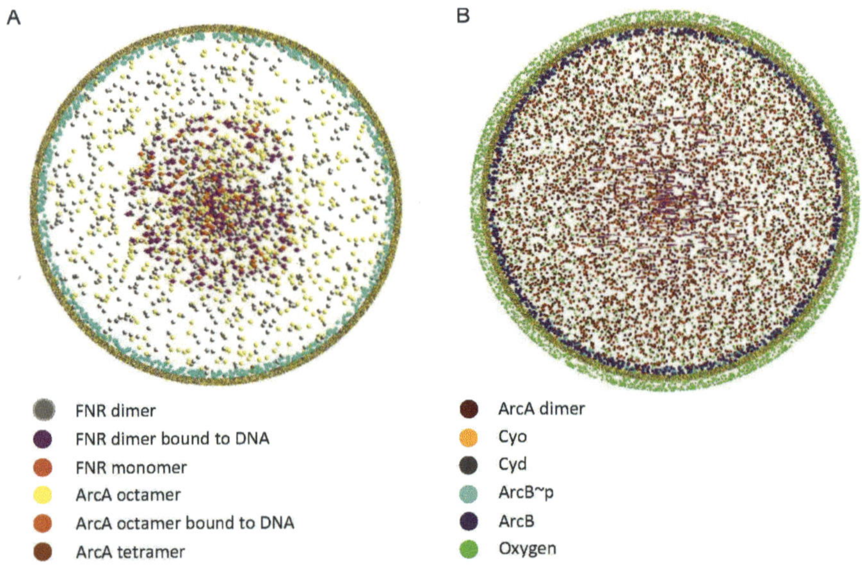

Katja Bettenbrock *et al.*, Figure 2.9 The initial and final states of model with no oxygen and with sufficient oxygen supplied. (A) The initial state of model with no oxygen supplied to the cell. (B) The final state of model with sufficient oxygen (217% AU) supplied to the cell. Each type of molecule is represented by a different colour as shown.

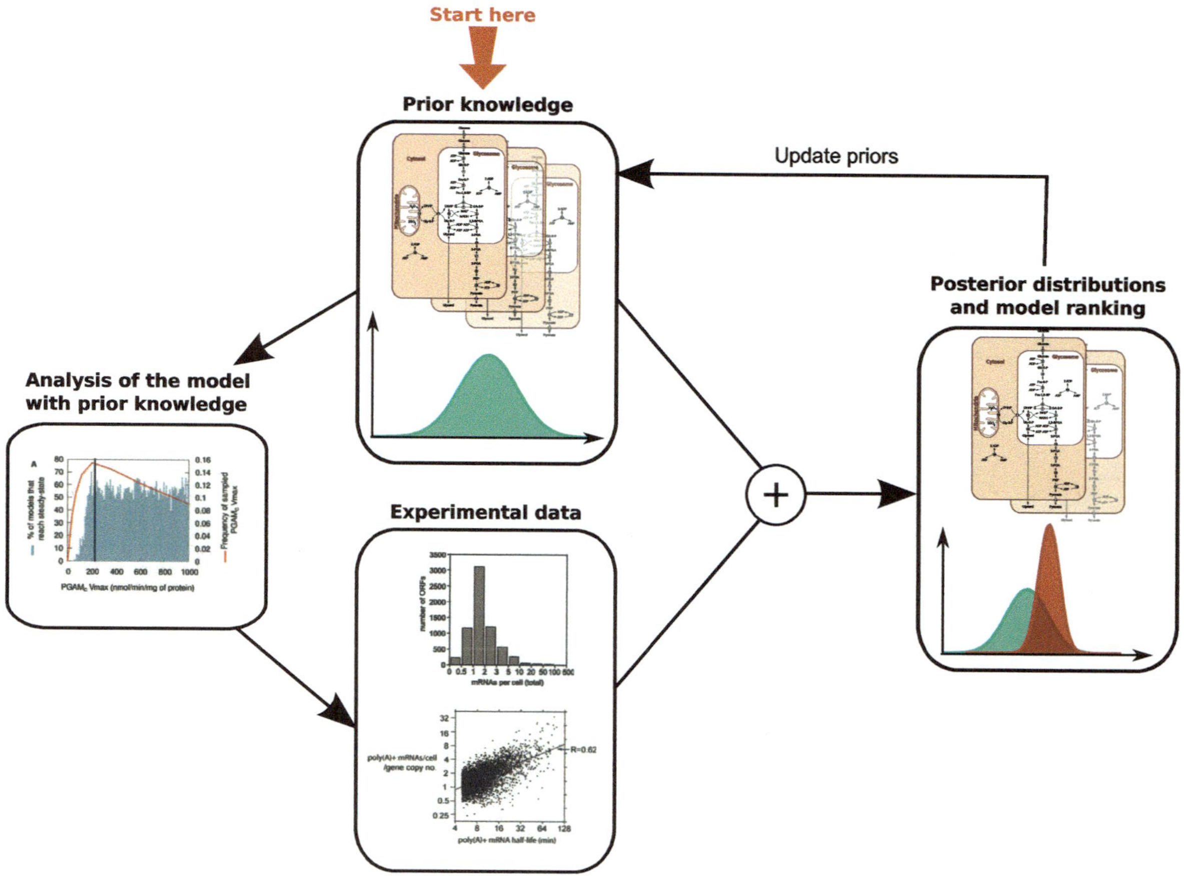

Fiona Achcar *et al.*, Figure 3.1 General Bayesian framework for explicitly handling uncertainty in the SilicoTryp project.

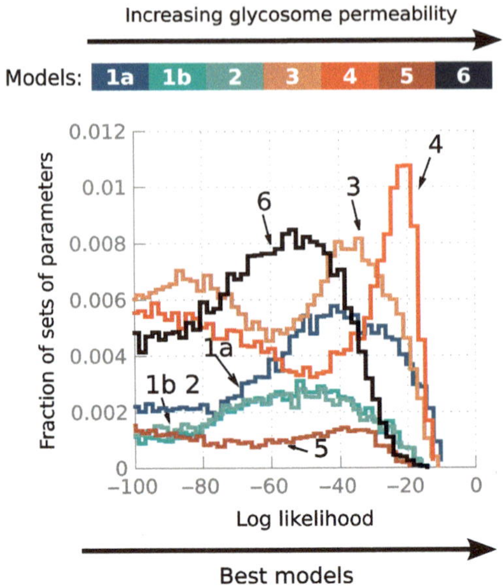

Fiona Achcar *et al.*, Figure 3.5 Log-likelihood (goodness of fit to experimental data) of models with increasing glycosome permeability and cytosolic activities of the glycosomal enzymes. The higher the log-likelihood, the better the match between simulations and experimental results. Each model is simulated with a range of parameter sets that describes our uncertainty about the parameter values. See Achcar et al. (2013) for details. *Modified with permission from Achcar et al. (2013). Copyright 2013 John Wiley and Sons.*

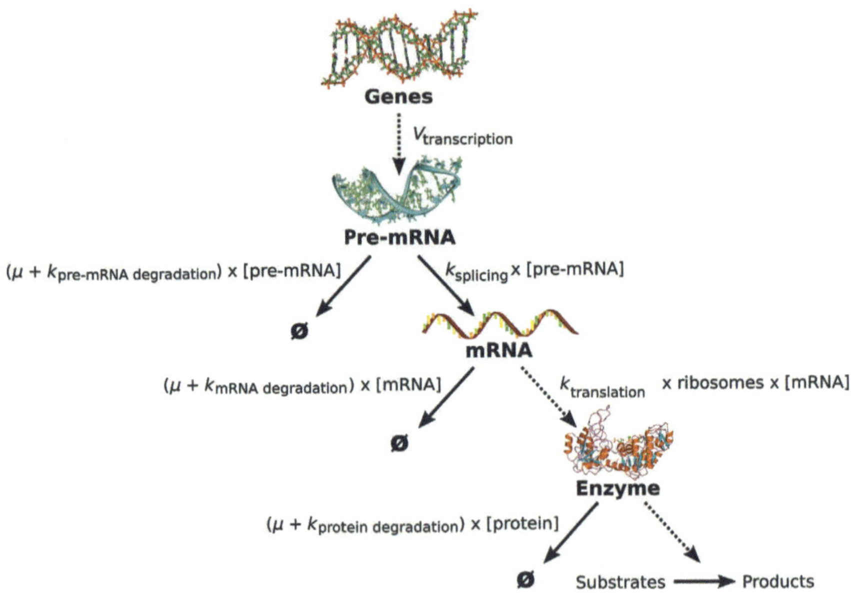

Fiona Achcar *et al.*, Figure 3.9 Transcription/translation module that will be added for every enzyme of the model. The module is based on the model published by Haanstra, Stewart, et al. (2008). The rate constants of the various processes involved are indicated by k, and μ are the specific growth rate of the trypanosomes, and *ribosomes* represents the number of ribosomes per molecule of mRNA.

FSC
www.fsc.org
MIX
Papier | Fördert
gute Waldnutzung
FSC® C083411

Kolibri 360 GmbH
Ferdinand-Jühlke-Straße 7
99095 Erfurt, Deutschland
produktsicherheit@kolibri360.de